Technology Dynamics

Technology Dynamics

The Generation of Innovative Ideas and Their Transformation into New Technologies

Angelo Bonomi

CRC Press is an imprint of the
Taylor & Francis Group, an informa business

First edition published 2020
by CRC Press
6000 Broken Sound Parkway NW, Suite 300, Boca Raton, FL 33487-2742

and by CRC Press
2 Park Square, Milton Park, Abingdon, Oxon, OX14 4RN

CRC Press is an imprint of Taylor & Francis Group, LLC

International Standard Book Number-13: 978-0-367-43846-3 (Hardback)
International Standard Book Number-13: 978-0-367-42562-3 (pbk)
International Standard Book Number-13: 978-1-003-00613-8 (eBook)

CV 09.10.2020 1306

Dedication

To my wife Giacinta and my sons Marco and Pietro
A special thanks to Pietro for the drawing on the cover

Contents

Preface

The gestation of this book has a long history since I was a young researcher at the Geneva Research Center of the Battelle Memorial Institute. In this center, hundreds of R&D projects were carried out each year; however, a only few of them were continued or terminated with the entrance of a new technology into use. In fact, most of them were abandoned. I questioned myself about what happens to the huge amount of knowledge generated by the numerous abandoned projects that apparently passed into oblivion, why industry invested a lot of money in R&D projects, most of which were abandoned transformed into financial loss. Actually, I found answers to these questions much later, and I have tried to report them in this book. After about ten years of activity in R&D projects, mainly in the metallurgical and environmental fields, scientific research in the field of electrochemistry, my interests were directed to what happens to technologies after R&D development, and I obtained a transfer to the Department of Applied Economics, addressing studies on technology assessments and the influence of new technologies on markets of industrial products and equipment. However, after a number of years, I decided to join a spin-off of two my colleagues that had created a start-up in nearby France that consisted of in a laboratory for contract research and generation of start-ups. The laboratory succeeded in the creation of four start-ups, one lasted as an industrial company for more than 30 years, but after ten years of activity, the laboratory was closed due to lack of available venture capital for its projects. After this experience, I had a completely new idea: to start a consulting activity for promoting technology innovation in small firms, particularly in Italian industrial districts. To my surprise I found in these districts activity for technology innovation completely different from that which I had known during my activity at Battelle Geneva with the big industry. In fact, in industrial districts innovations were obtained as results of the creativity of people exploiting and combining existing technologies, practically without creating any real R&D activity, often imputing the investments in realizing innovations to maintenance or general company expenses. Nevertheless, the technological level of the firms was high, dominating their market niches; for example, the two districts in which I was involved in my activity composed of several hundred firms, were in fact, the greatest world producers of faucets and valves, falling just after China and before Germany with technological products of the highest level. During this period of activity, I became aware of a completely new point of view on understanding the innovation process. It happened through casual reading of a book, *The Complexity Advantage*, written by Susan Kelly and Mary Ann Allison, about the contribution of the science of complexity to increased business performance. I discovered through this book the activity of the Santa Fe Institute, along with many studied phenomena that could explain some open questions about the technology innovation process, particularly the nonlinear dynamics of technology, the behavior of networks, and the existence of cyclic triggered processes that could exhibit an autocatalytic expansion, or a slackening until their arrest. I considered these phenomena in this book for explanation of behavior of the structures for technology innovation. In particular I was interested in ideas on technology and technology innovation of two scholars of this institute: an economist, Brian Arthur, a theoretical biologist, Stuart Kauffman. Their ideas on the nature and evolution of technology, and the fact that a technology may be seen as a set of operations similar to a set of genes in biology, is at the base of my study on technology dynamics. Becoming a member of the association La Storia nel Futuro, which is active in the organization of study tours for Italian students in the Silicon Valley, I had the occasion to discover, through some study tours, another territory with ideas about innovation completely different from the traditional industrial views, with some similarities, but also important differences from those of Italian industrial districts. In this territory, innovators seem to have the idea that return of investments and capital values are simply the fatal by-products of innovation; researchers consider science to be just a means to obtain new technologies, and fundamental research is considered to be an activity of high interest that soon or later will be a source of highly competitive radical technologies. Comparing the innovative systems of Italian industrial districts with the traditional industrial innovative system of Silicon Valley with respect to the development of new technologies has been of great importance in understanding some fundamental aspects of the technological innovation process. This comparison gave me the idea to include all these studies and experiences in the description of a technology dynamics based on technological processes that may be explained by a suitable model of technology, and on the existence of organizational structures for technology innovation, such as the industrial R&D project system and the start-up – venture capital system, to which may be added the nascent industrial platform system, already used in informatic and communication technologies, now diffusing into implementation of these technologies in the manufacturing industry. Later in my career, I had the opportunity to have an external collaboration with IRCrES, the Research Institute on Sustainable Economic Growth of the National Research Council of Italy, which opened to me easy access to the vast literature in the field of innovation and its relation to economics, and, in turn, gave me great help in developing my idea of technology dynamics. However, a minor part of this literature appeared to me, through my scientific and technical education, of poor relevance and in contrast to the Galilean principle that *our discussion shall be on the real world and not on the world of paper.* All these considerations motivated me to write a book in which the experience in technology innovation is relevant, in which technology is discussed in the phase in which it is generated, rather than when it has its economic effects, a book enriched by real examples, perhaps sometimes of minor technological and economic importance, but useful for explaining technological concepts and dynamics of innovation.

Angelo Bonomi

Acknowledgments

There are many people who have contributed to the writing of this book. First of all, I cite Sergio Pizzini, former professor of physical chemistry at the University of Milan, who taught me how to do scientific research, as well as the relation of scientific research to R&D, and later encouraged me to write this book and provided help in its publication; Georges Haour, my colleague at Battelle Geneva Research Centre, and afterward, Professor of Technology Management at IMD, a business school based in Lausanne (Switzerland), for many discussions and suggestions during my studies on technology innovation and for writing this book; Emilio Sassone Corsi, editor of the web review *Management Innovation News*, for the discussions held during the writing of this book, particularly about crowdfunding of start-ups; Mario Marchisio, PhD in physics from the University of Trento, now Associate professor in synthetic biology at the University of Tianjin, People's Republic of China, for his help in the mathematical development of the model of technology and useful discussion on technological innovations in the field of synthetic biology; Paul Hobcraft, business consultant, who has introduced me to the world of industrial platforms. Finally, I thank Paolo Marenco, President of the association La Storia nel Futuro, for giving me contacts with universities and companies during my study tours in the Silicon Valley; and Secondo Rolfo, Director of IRCrES, for having offered me a position of senior research associate in his institute, despite my virtual lack of academic experience.

Author Biography

Angelo Bonomi obtained a Doctor in Industrial Chemistry degree from the University of Milan in 1969, and started his career in 1970 as a researcher at the Geneva Research Centre of the Battelle Memorial Institute, carrying out R&D projects, becoming head of the extractive metallurgy section, scientific researcher in the field of electrochemistry, and a member of The Electrochemical Society (New Jersey). In 1980, he changed his activity, collaborating with the Applied Economics department of Battelle, conducting studies on technology assessment and the impact of technology innovations on industrial products and equipment. Later, in 1988, he joined a spin-off of his colleagues, becoming managing director of Extramet, a French contract research laboratory and generator of start-ups. In 1993, he started an activity as consultant for environmental technologies, particularly for technological innovation in Italian industrial districts. His major contribution was the foundation of Consortium Ruvaris for the cooperation in R&D of about twenty firms producing faucets and valves. From 2001 to 2004, he taught management of technological innovation for a Master's in Engineering Management program at SUPSI, a technical university in the Italian-speaking region of Switzerland. Since 2013, he collaborates as Senior Research Associate at IRCrES, the Research Institute for a Sustainable Economic Growth of CNR, the Italian National Research Council, conducting studies on technology innovation and territorial innovation systems. As a board member of the association La Storia nel Futuro, which is active in organizing study tours in the Silicon Valley for Italian students since 2005, he has also participated to several tours for the study of the innovation system in the Silicon Valley.

Abbreviations

BDC	Battelle Development Corporation
BRIC	Brazil, Russia, India, and China
EU	Europe Union
GCC	Gulf Cooperation Council
GDP	Gross Domestic Product
GERD	Gross Domestic Expenditure on R&D
ICT	Information and Communication Technologies
ILC	Innovate-Leverage-Componentize
ISE	Innovative System Efficiency
KAIS	Korean Advanced Institute of Science
KAIST	Korean Advanced Institute of Science and Technology
KIST	Korean Institute for Science and Technology
LbyD	Learning by Doing
MIT	Massachusetts Institute of Technology
MVP	Minimum Viable Product
OECD	Organization for Economic Cooperation and Development
PARC	Palo Alto Research Centre
PC	Personal computer
R&D	Research and Development
RDK	General R&D Knowledge
ROI	Return of Investment
SME	Small and Medium Enterprise
SVC	Start-up – Venture Capital
SVIEC	Silicon Valley Italian Executive Council
VC	Venture Capital

1 Introduction

This book concerns the description of the dynamics of technological activities with knowledge that may be useful for the generation of innovations and policies for economic development and prosperity. Nevertheless, this book shall not be considered a text on management of technology innovations; in fact, it principally describes the dynamics of the formation of new technologies and not the effects of new technologies in the socio-economic system. Actually, technology innovation does not have any fundamental theory based on a general model of technology that makes possible the explanation of technological processes occurring in organizational structures such as R&D and startup-venture capital (SVC) systems, as well as in the nascent industrial platform system. The knowledge of dynamics of processes and structures of innovation may contribute to those theories useful for the management and promotion of innovation. Furthermore, knowledge of the dynamics of technology might be complementary to other disciplines, such as the economy of innovation, Of course, technology innovation is an extremely complex field and we are aware that our description of technology dynamics cannot be an exhaustive vision of this subject, but it may contribute to better understanding. The arguments about innovation are treated from a technological point of view, and not through a scientific or economic vision. This means that technology dynamics does not enter discussions about the complex relations and models existing between technology innovation and business or economic activities. Rather, it considers the existence of an important interface between technology and both scientific research and economic reality, but limits the study to technological factors that directly influence scientific, economic, and business activities. Description of technology dynamics is therefore not based on scientific or economic considerations, but on taking inspiration from analogies between biological evolution and technological evolution despite the existence of some important differences between them. This analogy was observed for the first time by George Basalla, professor of history of technology at the University of Delaware, and was reported in his book on the evolution of technology [1]. Technology dynamics mainly conceives innovation through processes that are apt for its realization, and not through the relations that innovation has with scientific or economic environments; it therefore sees organizational structures for innovations in terms of existing relations and behaviors internal to a technology. On the other hand, if there is vast literature on the study of technology innovation activities in relation to scientific research, and even more on R&D effects on economic growth, while sources of funding, performances, incentives, and motivations for technology innovation activities are reasonably well considered by academics and policymakers, the complex process by which scientific results are exploited and transformed in new technologies through an innovation process is poorly documented and studied little [2]. For these reasons, in writing this book, only a minor part of the literature on technological innovation related to economic and business activities has been taken into consideration in the study of technology dynamics; only what appeared coherent with the reality of a long experience in technology innovation was selected. A particular aspect considered in writing this book is the contribution to the study of certain territorial innovation systems that present a dynamics of technology innovation that is largely different from that of typical industrial R&D activities. These systems occur in the case of the Italian industrial districts and in the case of the Silicon Valley. Many fundamental aspects of technology dynamics have been identified by comparing these two systems with the traditional industrial system for the development of new technologies. The aim of this book is to describe the complex activity of technology innovation starting from generation of innovative ideas and their transformation into new technologies, and it is based on the idea that technology evolves continuously with time, and is changed by innovations characterized by a dynamics constituted by technological processes occurring in organizational structures such as R&D, the SVC system, and the nascent system of industrial platforms. Furthermore, it has been considered that an important source of new technologies is also constituted by the use of technologies particularly through learning by doing (LbyD), defined as an activity, increasing manufacturing, designing and use experience, leading to a positive macroeconomic production externality independently of bringing additional capital or work, and even R&D investments [3]. In the study of technology dynamics, various concepts and models are used that are the result of an innovative approach useful in developing descriptions of the reality of innovation. The validity of the described processes and structures is represented by their potential to explain innovative phenomena coherently with the many real examples of histories of technology innovations reported in the literature or experienced directly by the author. The study of technology dynamics is fundamentally based on a suitable definition of technology and a general model of technology that is able to explain: the various technological processes [4], the various organizational structures cited previously, such as that of R&D projects activity [5], the SVC system [6], and the industrial platforms system [7]. These structures are the result of an evolution that started in Germany around 1870 with the R&D activities in the field of dyes [1], and further evolved in the second half of the twentieth century with the entry of numerous types of actors, making R&D other than in industrial laboratories into what has been called by Georges Haour the system of *distributed innovation* [8], enriched with new models of business in what it has been called by Henry Chesbrough an *open innovation* regime [9]. Such evolution at the beginning of the twenty-first century has also given birth to a new form of relations between industrial demand and offers of technologies in the context of what has been called industrial platform, finding its diffusion in the applications of ICT in the manufacturing industry [7]. However, all these organizational structures shall not be considered alternative, but rather, evolutionary. Actually, R&D is present in SVC activity and startups may be included in the structure of an industrial platform. In fact, all these organizational structures have been formed in response to needs for the development of various types of technologies and the existence of different strategies of innovation financing. In the description of processes and structures of technology innovation, we have made reference to ideas, models and real cases reported in the literature that confirm the reality of our descriptions; however, we have not tried to search and to report all existing literature that is coherent with our ideas, limiting us to a few references that we have considered as valid examples for explaining processes and structures of technology dynamics. The models described in this book concerning technology and the simulation of R&D and SVC activities may have a mathematical description useful for the study of the models. However, in the text of the book we have tried to give an exhaustive but qualitative description without use of mathematical expressions, which are reported instead in the appendices of the book for readers who may be interested in these aspects of the models. Finally, it should be noted that in the study of technology dynamics, despite the strong relation that exists between science and technology, we have made a distinction between scientific activity and technology innovation activity, considering that, in our opinion, corresponding processes and organizational structures influencing technology dynamics are different, resulting in differences in various aspects of management of these two activities, taking into consideration at the same time the relation of technology to science following the views of Bruno Latour with his definition of *technoscience* [10].

After this introductory chapter, this book includes four other main chapters. The second chapter discusses a general definition of technology and the origin of new technologies based on combinatory processes exploiting or not scientific results. The adopted model of technology is then explained, along with its use in the description of processes and steps composing the technology innovation activity from the generation of the innovative idea to the use of the new technology. We then further discuss the processes of transfer of technologies associated with a definition of know-how, and the relation of the model of technology with patents. The third chapter describes the three organizational structures: Research and Development (R&D) systems, Start-up Venture Capital (SVC) systems with their respective simulation models and industrial platform systems, followed by the comparative study of these three organizational structures. The fourth chapter discusses some applications of technology dynamics concerning the description of the various aspects of the innovation process, applications in statistical studies, particularly on research and innovation studies, on the relation between R&D investments and growth, and on studies based on patents. Finally, suggestions are presented, derived by the study technology dynamics, about the promotion of technology innovations. The book concludes with a fifth chapter, which describes the main results that have been found in technology dynamics studies.

REFERENCES

[1] Basalla G. 1988, *The Evolution of Technology*, Cambridge University Press, Cambridge, UK.

[2] Auerswald P.E., Branscomb L.M. 2003, Valley of Death and Darwinian Seas: financing the invention to innovation transition in the United States, *Journal of Technology Transfer*, 28, 227–239.

[3] Arrow K.J. 1962, The economic implications of learning by doing, *Review of Economic Studies*, 29, 155–173.
[4] Bonomi A., Marchisio M. 2016, Technology modelling and technology innovation: how a technology model may be useful in studying the innovation process, *IRCrES Working Paper*, 3/2016.
[5] Bonomi A. 2017, A technological model of R&D process, and its implications with scientific research and socio-economic activities, *IRCrES Working Paper*, 2/2017.
[6] Bonomi A. 2019, The start-up venture capital innovation system: comparison with industrially financed R&D projects system, *IRCrES Working Paper*, 2/2019.
[7] Bonomi A. 2018, I Canali Innovativi di Industria 4.0 e le PMI, *IRCrES Working Paper*, 7/2018.
[8] Haour G. 2004, *Resolving the innovation paradox: enhancing growth in technology company*, Palgrave Macmillan, New York.
[9] Chesbrough H.W. 2003, *Open innovation: the new imperative for creating and profiting from technology*, Harvard Business School Press, Boston.
[10] Latour B. 1987, *Science in action*, Harvard University Press, Cambridge, MA.

2 Technology

2.1 DEFINITION OF TECHNOLOGY

The activity that has existed since the dawn of humanity and has developed through the production of artefacts, products, and services, is built around a concept that is not clearly defined and carries out a whole range of meanings from the Greek original word *techne*, translated as *technique* in English and French, *Technik* in German, and *tecnica* in Italian, and typically used in philosophical discussions [1]. English word *technology*, *technologie* in French, *Technologie* in German and *tecnologia* in Italian, are formed by a combination of Greek words, *techne* and *logos*, meaning originally discussion about the *technique* and now largely used by people involved in this activity, either in a general context or in reference to a specific technology. When speaking about technology dynamics, we intend the word *technology* to be defined more accurately: *an ecosystem that includes all the specific technologies with their interactions and evolution.*

There are many operational definitions given to technology, such as the production of artefacts and services, or a collection of devices and practices available to a culture, and often, an application of science for the production of goods and services. Actually, the more useful definition of technology, for the purpose of the study of technology dynamics, is the general definition proposed by Brian Arthur in his book on the nature of technology [2], simply, *a means to fulfil a human purpose.* There is another aspect of technology that is of interest, considering the human purposes in using a technology. In fact, a technological activity may be seen as a set of physical, chemical, and biological phenomena in action that produce some expected effects that may be considered independently of the human purposes for their exploitation. A corollary of this view is the neutral nature of technology; the benefitting or dangerous aspects of its effects depend on the means and purpose of its use by humans, and not by the nature of the technology. For example, bows and arrows may be used to kill prey to ensure survival, or to kill a man during a fight. Also, if it is true that very often a technology is developed for a specific purpose, that does not limit the fact that it may be used for another completely different purpose. On the other hand, considering technology as an exploitable effect of a set of physical, chemical, and biological phenomena, it is easy to understand the importance of science in the generation of new technologies.

There is also a diffused use of the word *technology* to indicate a certain type of technologies that do not have technological, but socio-economic, objectives. These technologies, often also called *new technologies*, are normally based on the use of information and communication technologies (ICT) that are able to typically develop new services in the social and economic fields, using capabilities such as: big data storage, cloud computing, tools such as computers or smartphones, and infrastructures such as Wi-Fi and internet. These technologies are generated typically by a combination of informatic and communication expertise with new ideas for possible social, commercial, or economic applications. Applications include financial and banking activities, social networks, e-commerce, marketing, and social behaviour. There are similarities, but also many differences with innovations with technological purposes, and different criteria, not of technical nature, such as the Minimum Viable Product (MVP) i.e., the minimum development that shall be reached to start customers market testing, and the possibility of pivoting the start-up objective if the market target is not reached. Furthermore, these types of technologies are not normally related to scientific research. Although these types of technologies have reached a great importance from social and economic perspectives, they are not taken into consideration in the description of processes and structures of technology dynamics in this book, because of the great differences in their development process and in their objectives that are socio-economic, rather than technological in nature. On the other hand, it should be noted that these new technologies with socio-economic objectives depend basically on innovations made with technological objectives that determine their shape and operative possibilities.

2.2 GENERATION OF NEW TECHNOLOGIES

Going further in the description of technology, we observe that technology may be considered as a structure formed by a set of components of a technological artefact, or be based on technological operations necessary to produce or use such an artfact. At the same time, we may consider that in a new technology there are always present pre-existing technologies composed by previous used types of operations or components. This fact has lead Brian Arthur in his book on the nature of technology [2] to consider that *a new technology is composed of a combination of pre-existent technologies able to exploit new phenomena discovered by science.* An example of generation of technology of this type is laser, which may be considered to be formed by an optical cavity obtained with mirrors and an electronic oscillator generating a flux of light. These pre-existing technological components are arranged in such a way that makes exploitation of the quantum phenomenon of coherent emission of light possible. Actually, looking to the various technologies developed during the evolution of this activity, we may also observe that there are important technologies that have, in fact, been generated by a simple combination of pre-existent technologies without a direct exploitation of phenomena discovered by science, although such phenomena have been possibly exploited in the components or operations composing the new technologies [3]. Such types of generation of new technologies are diffused, for example, in mechanical or electronic designs of new products that in fact do not exploit any new scientific discovery. Thus, a complete definition of the generation of technologies may be: *a new technology is composed of a combination of pre-existent technologies able or unable to exploit new phenomena discovered by science.* The difference between technologies generated by exploiting or not exploiting scientific discoveries is of great importance in technology dynamics as it leads to important differences in the process of generation of ideas and their transformation into new technologies. In order to illustrate this fact well, it is useful to present some real examples of important technologies with these different origins. For this purpose, we have chosen the case of invention of the photocopier by Chester Carlson, in which the discovered phenomenon of electric charge of materials by light has been exploited, and two other examples of technologies obtained by simple combinations of previous technologies, such as the invention of the personal computer (PC) by Steve Wozniak and the invention of the coffee maker Moka Express® by Alfonso Bialetti; this last invention resulted from observations made in a technological field completely different from those of other coffee makers. The history of these technologies is described as follows:

> Photocopy was invented by Chester Carlson in the thirties of the past century and development financed by the Battelle Development Corporation (BDC), a division of the Battelle Memorial Institute [4]. His central idea was to exploit the photoelectric phenomena existing for certain materials, in form of photoconductive film, exposed to light in such a manner to reproduce, for difference of charges, an image attiring fine carbon powders that may be used to print a paper page. Photoconductive properties of materials were discovered in the last decades of XIX century and Chester Carlson was probably aware about these phenomena during his studies in physics at the California Institute of Technology. He made experiments in his own kitchen with good results sufficient to obtain a valid patent in 1937. After a period of interruption because of the war, in 1944 Carlson signed an agreement with the BDC, for the development of the invention by R&D activity in the Battelle Columbus Laboratories. At the end of 1946 Battelle was in measure to make an agreement with Haloid, a medium sized company in the field of photographic paper, for the development and industrialization of the invention. At the end of the fifties Haloid succeeded in offering a first automated model with a strong market development and becoming the present Xerox company.

The PC may be considered a typical combinatory innovation without any direct exploitation of scientific phenomena. Its origin and development were the results of efforts of many people and companies; however it is usual to cite the pioneering role of Apple and its founders Steve Wozniak and Steve Jobs described in the official biography of Steve Jobs written by Walter Isaacson [5].

> The invention of PC may be attributed to Steve Wozniak that was at that time an electrical engineer working at HP on connection of a terminal constituted by a keyboard and a monitor with a central minicomputer. Hobbyist in electronics, he frequented the Homebrew Computer Club. In one of meetings of this club there was a discussion on microprocessor, Incidentally, the microprocessor is an electronic device for computers, not exploiting directly any new scientific discovery, invented by Federico Faggin, an Italian physicist working at Intel, a company spin-off of Fairchild Semiconductors. Wozniak had the idea to use the microprocessor to put in the terminal itself some capacities of the minicomputer, making a stand-alone computer on a desktop, in fact a PC. Immediately Wozniak worked on realization

of needed circuits succeeding to connect a keyboard input giving a wanted output on a TV screen on Sunday, June 29, 1975, a milestone for PC. After that, with his friend Steve Jobs, founded Apple in 1976. The first product called Apple 1 was simply a motherboard, connected to a typical keyboard, used also in electric typewriter, and to a domestic TV apparatus. Steve Jobs may be considered the person that understood fully the potentiality of Wozniak machine as a product, easy to use, inexpensive, interesting people in general and not only professionals or hobbyists. In fact, before Apple there were other desk computers, the first being Olivetti P101 in 1964, invention that exploited for the first time magneto-striction phenomena for the memory with the reduction of the storage volume necessary for a use as desk computer, and being in this way a new technology depending from exploitation of a discovered scientific phenomenon. This model used a printer, and not a monitor, for input of data or output of results, it will be followed by a more advanced model the HP 9100 in 1968. However, both models were relatively expensive and addressed to professionals. In the case of Apple innovation components were arranged following a functional computer structure called Von Neumann architecture, known since 1944. Exploitation of new phenomena had been present only in used commercial components, such as for example the use of transistor effect discovered in 1925 and the possibility to use silicon as solid transistor discovered in 1948.

In addition to the PC as an example of combinatory innovation, we report another radical combinatory invention as an example of a typical technological innovation of Italian industrial districts: a case of apparent paradox of innovations of certain importance in small firms, but not related to any scientific research activity [6]. That is the case of Moka Express®, a coffee maker in competition with a pre-existing coffee maker called Napoletana. Their different design concepts are illustrated in Figure 2.1. Napoletana coffee maker has a bottom containing water for boiling and an upper filter containing coffee powder; when water is boiling, the coffeemaker is overturned from position 1 to position 2, as reported in the figure, boiling water goes through the filter to be collected at the bottom of the overturned coffeemaker, and coffee served. In the Moka Express®, boiling water is prepared in the bottom part of the coffee maker and the steam goes through the filter containing coffee powder, coming out as coffee from the upper tube. Details on generation of this new coffeemaker are given as follows:

Moka Express® was invented by Alfonso Bialetti. He emigrated in France at the beginning of the XX century and came back to Italy in 1918 with experience in aluminium casting opening a small mechanical workshop. He invented the new coffeemaker at the beginning of thirties starting production in 1934. It is remarkable that Moka Express® design was not derived by a new combination of elements of other existing coffeemakers but by a pot used in washing laundry in which boiling water comes through a tube from separated heated bottom of the pot. Differences from Napoletana coffeemaker was not only in design but also in material, aluminium instead of copper sheet, fabrication, aluminium pressure molding instead of welding. After the war his son Renato Bialetti developed the product with a successful marketing effort expanding sales not only in Italy but also abroad while production of Napoletana coffeemaker practically disappeared.

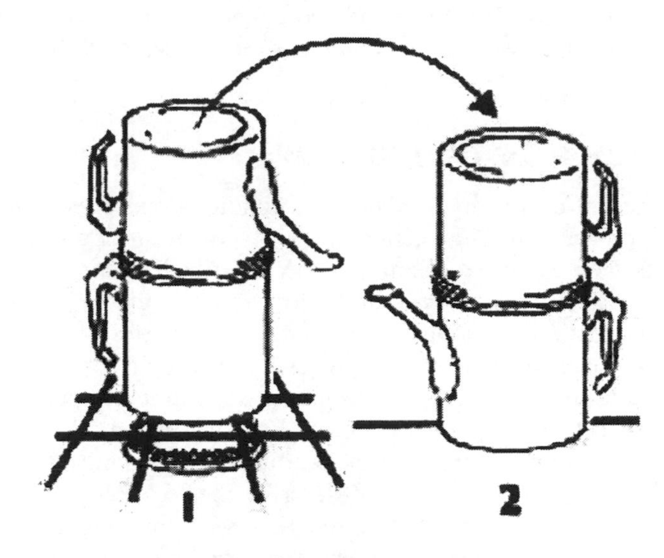

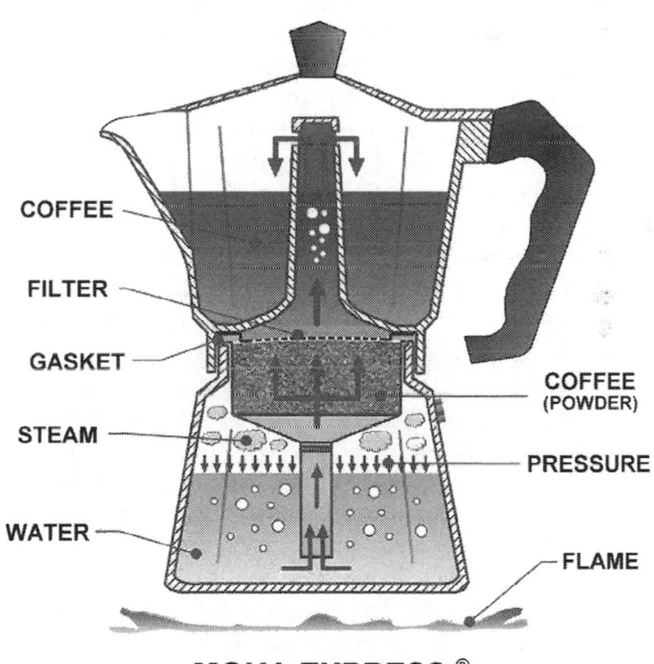

FIGURE 2.1 Coffeemakers Napoletana and Moka Express®.

Moka Express® may also be considered a good example of radical combinatory development based on technologies not necessarily belonging to the same technological sector.

2.3 THE MODEL OF TECHNOLOGY

The development of a general model of technology may have two approaches. The first approach considers technology in terms of an artefact formed by its components. For example, a car may be described as a set of components including motor, brakes, wheels, and tires, and the model considers this set and the interactions among components and performance. The second approach describes technology as a process formed by a set of operations, and the model considers this set and the interactions among operations and performance. In fact, the approach of technology as a process is broader because it may include technologies, such as production of chemicals, for which the characteristic of the product cannot describe the technology with which it is produced. On the other hand, technologies corresponding to the use of an artefact cannot be described in terms of components, but rather, in term of operations. Finally, it is also possible to describe artefact components through their assembly process. For these reasons, our model describes technology as a process composed by a set of operations that would, of course, be different if we consider making or using the same artefact. For example, we may distinguish the technology for making an airplane from that of flighting an airplane. In certain cases, a technology is not referred to a specific artefact; in the case of surgery, many types of artefacts may be used for its operations. It should be noted that technological operations have also the nature of technologies and, in this way, a set of operations representing a new technology may also be considered the result of a combination of pre-existent technologies. This has been previously considered in defining the formation of a new technology.

The model of technology that we use was born from ideas developed in the early 1990s at the Santa Fe Institute. This Institute, dedicated to the transdisciplinary science of complexity, was created in 1986, its founders were: George Cowan, former scientist at Los Alamos National Laboratories and first President of the Institute, Murray Gell-Mann, winner of the Nobel Prize in Physics, as well as many supporters, particularly Kenneth Arrow, winner of the Nobel Prize in Economics. Among the first fellows of this Institute was Brian Arthur, an economist, well known for his studies on the existence of increasing economic returns, and at that time, professor at the University of Stanford, Stuart Kauffman, a theoretical biologist, well known for mathematical modeling of gene interactions, at that time professor at the University of Pennsylvania. A discussion about technology between these two scholars at the end of the 1980s at the Santa Fe Institute is, in fact, at the origin of the model reported in a book describing the foundation and the main ideas characterizing the Santa Fe Institute [7]. The discussion started on the nature of technological change and Brian Arthur observed that economists do not have any fundamental theory and have treated technology as if generated from nothing, falling from the sky in the form of projects such as production of steel or fabrication of silicon chips. In fact, continued Brian Arthur, in the past, technology was not considered as part of the economy but an exogenous factor. More recently, there was the tendency to build up models of technology endogenously produced by the economic system as result of investments in R&D and technologies considered as any other goods. Brian Arthur thought that this view was not completely erroneous, but that was not the core of the problem. Considering the history of technology, it does not resemble a good; in fact, technologies do not come from nothing, but are often prepared by previous technological innovations and technology may be better considered as an ecosystem in evolution. Stuart Kauffman argued that technologies form strongly interconnected, dynamic, and instable networks. Such networks may present explosions of creativity and mass extinctions as in biological ecosystems. In fact, human transportation by carriages was replaced by the use of cars. At the same time, stations for changing horses were replaced by gas stations and there was the emergence of new industries for production of gasoline and tires. Brian Arthur observed that such processes are a good example of his concept of increasing returns as a new technology may create new niches for goods and services; he asked to Kauffman to try developing a model in which technology is activated at the moment of its creation, rather than appearing at the moment in which its effects are observed. That opened the idea of treating a technology mathematically as a set of operations, similar to a set of genes operating in a biological entity, and considering technological mutation similar to that of the origin of life, a research field in which Stuart Kauffman was active for about fifteen years. Following this discussion, Brian Arthur continued to study the core aspects of technology, later developing the idea that technology is the result of a combinatory process of previous technologies able to exploit new scientific discoveries [2]. On the other side Stuart Kauffman joined a team of researchers at the Santa Fe Institute to develop a mathematical model of technology considering technology as a process consisting of a set of technological operations [8].

2.4 THE STRUCTURE OF TECHNOLOGY

In the mathematical model visualized by Kauffman, technology is considered as a set of technological operations. Each operation is characterized by a set of instructions or parameters and each parameter is characterized by a set of choices or discrete values valid in certain range [8]. In fact, considering the sequence of the various operations over time, we may describe in a more detailed way a technology as a graph in which operations, either in sequence or in parallel, are represented by arcs oriented with time similar to that with which the various tasks are represented in the PERT method for project management. For example, considering a simple technology of heat treatment, we may have three operations: heating to a certain temperature, maintaining a certain time at that temperature, and cooling rapidly to room temperature. These operations may be represented in a simple graph structure of the technology in Figure 2.2. The parameters of

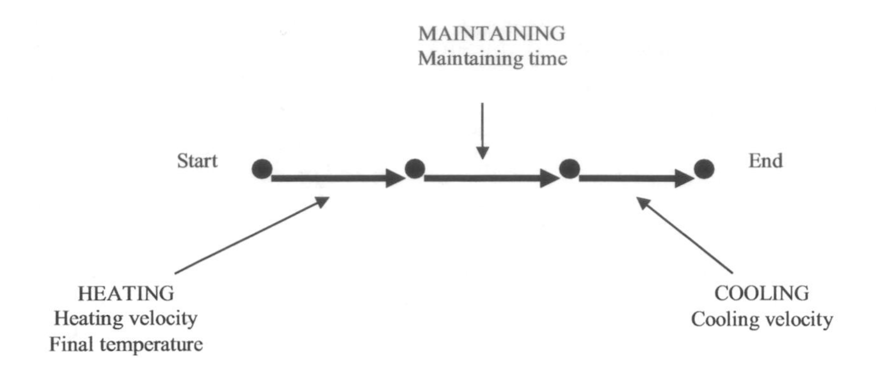

FIGURE 2.2 Graph model of a heat treatment technology.

the first operation would be temperature that shall be reached and the heating velocity, the parameter of the second operation would be the maintaining time, and the parameter of the third operation would be the cooling velocity. In order to complete the model of this technology, we have to establish for the first operation a certain number of values of temperatures and a certain number of values of heating velocities; for the second operation we have to establish a certain number of values of maintaining times; and for the third operation a certain number of values of cooling velocities, all of which taken within a specific range. The best conditions for the heat treatment would be obtained by the choice of the most valid values of these parameters. It shall be noted that the description of a technology structure with its operations may be established with more or fewer details following the purpose of the model. This is possible because operations themselves have the nature of a technology and may be described in terms of sub-operations. For example in Figure 2.3, we report the simplified structure of the technology of production of faucets, which are composed of mechanical components originating from bars and ingots of brass. Bars are hot-stamped and machined, while ingots are cast in components, followed by chromium plating. All parts are finally assembled to form the faucet. It shall be noted, as reported in Figure 2.3, that, for example, the chromium plating is in fact composed by the sub-operations of degreasing, nickeling and chroming.

The described structure of a technology with its set of operations, parameters, and parametric values may be treated mathematically using the so-called NK model. This model was used originally for modeling interactions among genes in biological entities [9]. In the case of the model of technology, genes are substituted by technological operations. Incidentally, it should be noted that the NK model has been also used in a mathematical model that considers technology as an artefact composed by a set of components [10]. Originally, the mathematical description of the technology model was used to reproduce the experience curve showing the decline of labor costs with cumulative

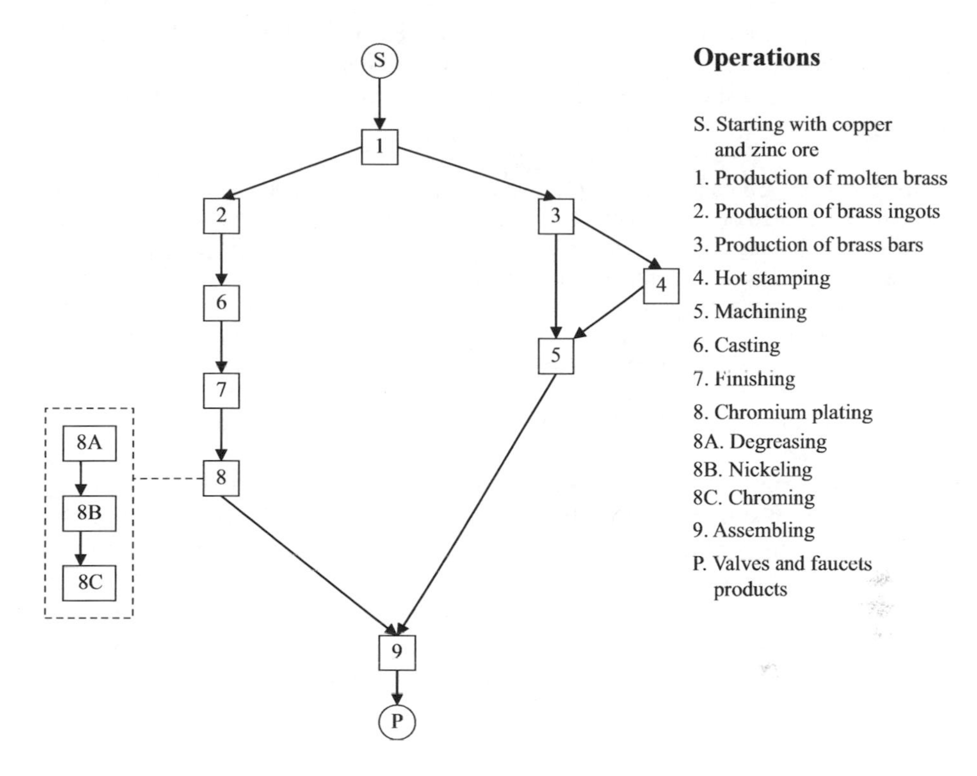

FIGURE 2.3 Graph model of technology of production of faucets.

production of a given manufactured good [8], observed at first in the airframe industry and expressed by the so-called Wright's law [11], resulting from the activity of LbyD. The mathematical model of technology is useful in describing various aspects of technology dynamics concerning concepts such as the technological space and landscape, as well as the space of technologies, and its full mathematical description is reported in Appendix 1 at the end of this book. We present next the main aspects of the model without using mathematical expressions.

2.4.1 Technological Space

The model considers, for each operation, the values or choices of parameters as a discrete set existing in a determined range. The whole set of established parameters values or choices of all operations corresponds to a *technological recipe* that may be considered in operating a technology [8]. Specific choice of parametric values for each operation constitutes then a configuration of the technology and, by a combinatory calculation, we can obtain the whole number of configurations or possible recipes existing for the modeled technology. All the configurations of a modeled technology may be represented mathematically in a multidimensional discrete space in which each point represents a specific recipe of the technology. Such space is called *technological space*. An idea about how it is built up a simple technological space is given in Figure 2.4 and explained as follows:

> In this figure are considered simply technologies with operations with one parameter that may assume only the values 1 or 0. If a technology is composed by only an operation its technological space will be represented by two points corresponding to 1 or 0. If it is composed by two operations there are four possible configurations and the technological space represented by four points in a square and, if it is composed by three operations, it will be consequently represented by eight points forming a cube. In the case of four operations the representation will be a hypercube actually a figure with four dimensions. If there are more parameters and values it will be necessary to add more dimensions to the technological space in order to represent all recipes of a technology.

In the technological space it is possible to measure the similarity of recipes by the Hamming distance between two points, or recipes, of the space. Hamming distance is defined in discrete mathematics as the minimum number of substitutions in the elements of a string to change the string into another of equal length. That corresponds in our case to the number of changes in values or choices that we introduce to make identical two technological recipes. Higher is the Hamming distance, lower is the similarity of recipes.

2.4.2 Technological Landscape

Considering that all recipes may be represented by points in technological space, we may attribute to each point or recipe a scalar value of efficiency of the corresponding recipe, obtaining, by mapping this space, a fitness landscape

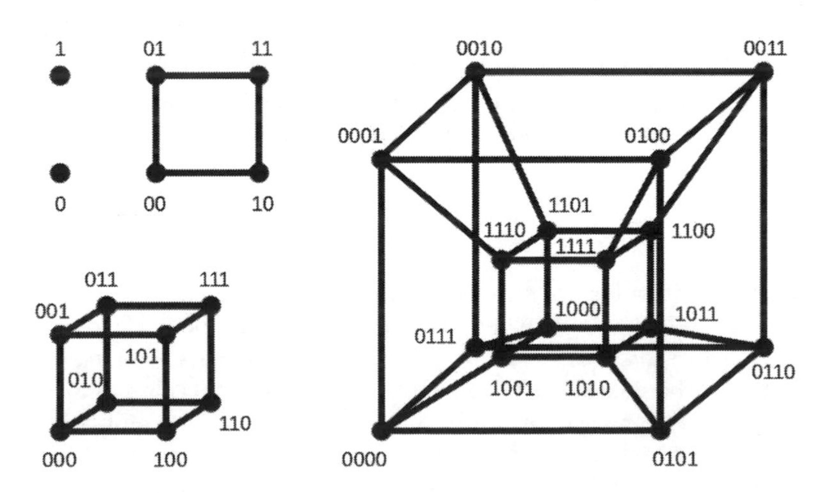

FIGURE 2.4 Representation of recipes in a technological space.

transforming in this way the technological space in what is called a *technological landscape* [8]. Such a landscape is characteristic of the specific structure of operations characterizing the modeled technology and the defined type of efficiency. From the perspective of the model, it is possible to define an overall efficiency of a specific recipe of a technology, but because of the technological nature of operations, to also define an efficiency of an operation with specific values for its parameters.

Technology efficiency (fitness) is a complex concept that is difficult to define quantitatively by a unique description. From a practical point of view, there are many types of efficiency that may be considered. For example, it is possible to consider energy efficiency of a technology in terms of production of energy, but on the contrary, also in terms of minimization of its consumption. It is also possible to define environmental efficiency of a technology in terms, of, for example, level of abated pollutants, as well as in terms of for example, level of purity and accuracy. For practical use of the model, it is useful to possibly choose a mode of calculation of efficiency in such a way that the calculated overall efficiency results from the sum of values concerning the efficiency of the various operations. For example, in a technology of production of energy there are operations that have a positive efficiency generating energy, as well as operations with negative efficiency consuming energy, and the overall efficiency may be calculated using the sum of positive and negative values related to efficiency of the various operations. One of the more important types of efficiency of a technology concerns its economy and may be expressed conveniently, considering the related cost of production transformed in terms of efficiency by inverting the total value of costs of all its operations. Actually, there are two ways to calculate the inverse of cost or efficiency of the technology, the first of which involves calculating the costs of operations and making the inverse of the sum of these costs; otherwise it is possible to also consider a different definition of efficiency of a recipe as average of the sum of efficiency of the single operations, i.e., considering the inverse of cost of each operation [8].

A technological landscape is characteristic of the modeled structure of a technology and may be different according to the type of technology efficiency considered, and, from the perspective of the model, the efficiency depends on the considered recipe. As the whole set of technology recipes is the result of a simple combinatory calculation, certain recipes would be absurd and have null or negative efficiency and others would have positive efficiency. Exploring a technological landscape, we may find regions with recipes with nearly null efficiency and other regions with recipes with high values up to optimum values of efficiency. The landscape may present in certain cases only an optimum of efficiency at the top of a single *hill* of the landscape or have cluster of *peaks* of efficiency or even a rugged structure of high number of *peaks* with roughly the same efficiency. In a technological landscape, the optimization of technology efficiency may be seen as an exploration searching of an optimal *peak* of efficiency for the technology. Actually, as a technology presents various types of efficiency with different landscapes, a practical optimization of a technology consists in reaching a good compromise among the various levels of efficiency of its landscapes. In Figure 2.5, we give a schematic view of a technological landscape showing a cluster with *peaks* of higher or lower recipe efficiency. In this figure, the multidimensional technological space has been simplified and points are arranged on a bidimensional surface for a three-dimensional graphic representation, in fact corresponding to a technology composed by one operation, with two parameters having two ranges of values. In a technological landscape, it is possible to mathematically study the search process of an optimal recipe for a technology in terms of exploration of the landscape, and there are many theoretical studies concerning such exploration or giving the shape of the landscape. For example, there is a study about the search of optimal conditions of efficiency [12], or discussing the search in terms of adaptive explorative walk [13], as well as a study on recombinant search in the invention process [14]. Technology landscapes have even been used, not necessarily as mathematical tools, in discussing certain aspects of technology management [15], and in technological searches in landscapes mapped by scientific knowledge [16]. This last study is interesting because it offers an explanation of how scientific knowledge may be used in improving a technology, but also in generation of new technologies without actually directly exploiting any new phenomena discovered by science. The mathematical study of the technological landscape has also given a contribution to the knowledge of the shapes assumed by the landscape. Such shapes depend on an effect called *intranality* that represents the influence

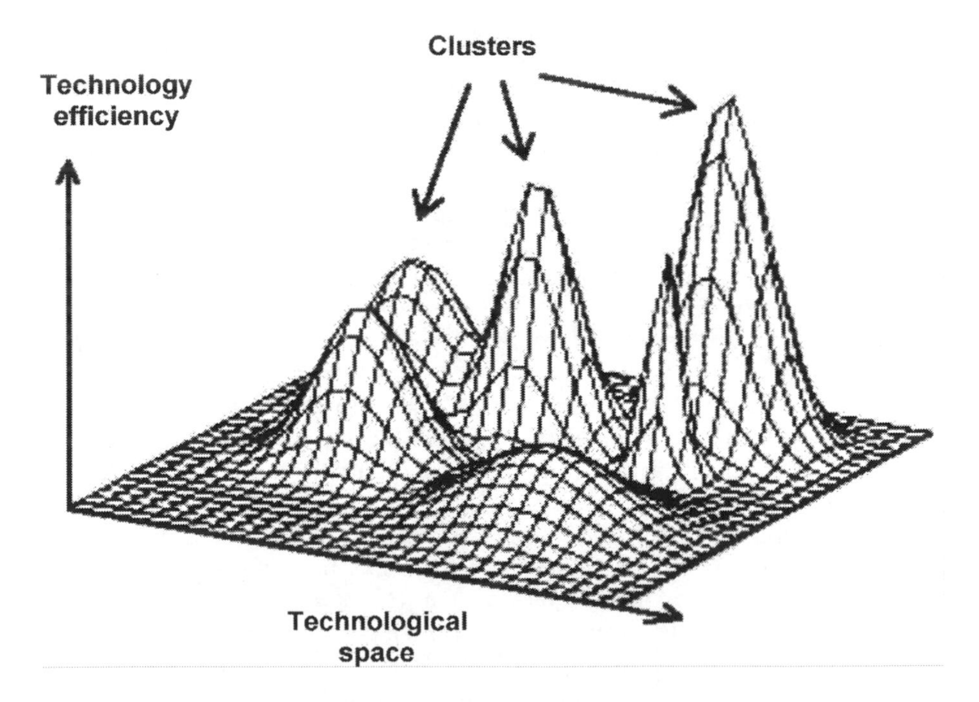

FIGURE 2.5 Schematic view of a technological landscape.

of changing the conditions of an operation, for example, by optimization of its efficiency, on the efficiency of other operations [8]. It is possible to show that: if the effects of intranality are absent, i.e., variations of conditions of an operation do not influence the efficiency of other operations, the landscape presents a single *peak* of efficiency; if there are some intranality effects on certain operations, the landscape presents a cluster of *peaks* of efficiency with various heights; if the intranality effects are very numerous among the operations, the landscape appears rugged with many *peaks* of similar height.

2.4.3 Space of Technologies

The technological space is useful to describe a single technology with a defined operations structure. However, when discussing various technologies, such as studying technological competition and evolution, it may be useful to have a space representing all considered technologies. Technology has been defined as an activity that is able to fulfil a specific human purpose [2], by consequence we can consider the existence of a set of technologies able to fulfil *the same* human purpose. It would be of interest to represent this set of technologies in a space in which it is possible to describe technology evolutions and evaluations of differences between technologies that are in competition for the same purpose. The structure of a technology can be described as a graph that may be represented mathematically by a corresponding matrix. A set of technologies having the same purpose may then correspond to a set of matrices. That opens the possibility of building up a multidimensional discrete space, considering the various combinations of the matrix elements analogously to the combination of configurations strings in the technological space, in which each point represents a matrix corresponding to a technology with a specific structure of operations. This representation is called *space of technologies* [3]. Such a matrix shall, of course, take account of all types of operations, present or not present in the technology, but included in all technologies having the same purpose and considered for a defined space of technologies. In this case, differently from the technological space, the Hamming distance among points is defined, comparing matrices, rather than strings of configurations. Such distance in the space of technologies increases with the difference between two technologies and may be considered a measure of the *radical degree* of a new technology compared to a pre-existing technology or alternative new technology. Following a largely used terminology, a technology may be considered by the model *radical*, if this distance is great, or *incremental*, if this distance is small. At the same time, a technological innovation may be considered radical (drastic) if the change necessary to transform a pre-existing technology into the new technology is great, or incremental (evolutive) if this change is small. In this way, the space of technology defined by the model offers a special view of what has been defined as natural trajectories of technical progress [17] in the frame of technological paradigms [18]. In this space, incremental technologies appear close to the pre-existing technologies, while a radical technology represents a transition at large distance with respect to the pre-existing technologies. It should be noted that the appearance of a new radical technology, that probably will have in its structures new operations not considered previously in the space of technologies, has as a consequence, an expansion of dimensions of the space of technologies in such a way that it may include the new radical technology. The radical degree of a new technology plays an important role in the resulting socio-economic impact of a new technology and on the probability of success of technology developments and influencing the various strategies that may be used for the development of technological innovations.

In conclusion, for the model of technology, the innovation is represented by a change of the structure of the technology, while the change of the operational conditions of a technology may represent only an improvement of the efficiency of the technology, resulting in an exploration of its technological landscape searching optimal conditions for the technology, typical activity of LbyD. However, it is not impossible that during the activity of LbyD, optimal conditions are found, not only by changing the operational conditions of a technology, but also by introducing minor changes in its structure, thereby realizing an incremental innovation. That explains the fact that technological innovations may appear, following the model of technology, as well as during the use of a technology, and not only as result of R&D activities. Finally, it should be noted that the space of technologies cannot represent the efficiency of the various technologies because efficiency depends on the recipe of a technology shown in its technological landscape, and thus it cannot be attributed to any scalar value of efficiency to the position of a technology in the space because it in fact depends on the specific recipe and on the type of efficiency represented by the landscape of the technology.

2.5 TECHNOLOGICAL PROCESSES

There are a certain number of technological processes that may be described using the model of technology. These are of two types: those occurring in the activity of a single technology and those occurring in the frame of a set of technologies. Other types of technological processes occur in specific organizational structures for innovation and they will be treated with the description of such structures. The technological processes occurring in a single technology are those due to externality and intranality effects. Those occurring in the frame of a set of technologies are the ramification process, the spandrel effects, the velocity of innovation, and a particular regime of technological competition called Red Queen regime. In addition to these processes, the important process of transfer of technologies in association with the procedural knowledge called *knowhow*, is described separately, followed by the relation existing between the model of technology and patents.

2.5.1 Externality Effects

During the use of a technology, there are many external factors that may influence its efficiency. These factors may include changes in various types of cost, changes in characteristics of raw materials, competition with other technologies, new regulations to be complied, changes in the requirements of the products, failure of components of equipment, etc. Such changes modify the form of the technological landscape, possibly reducing the efficiency of the used recipe. It is then necessary to search a new optimal recipe exploring the new landscape resulting from the externality effects, or even to change the structure of the technology to realize an innovation that is normally of an incremental type. The externality of a technology may also change the conditions of the compromise about the various optimums in the landscapes, corresponding to the various types of efficiency, obligating the search of a new compromise. The effects of externality are actually the major cause of the continuous normal process of evolution of a technology that may be observed with time.

2.5.2 Intranality Effects

The intranality effect, already cited in discussing the shape of technological landscapes, consists of the fact that changing the conditions of an operation in order to improve its efficiency may influence the efficiency of other operations of the technology [8]. Consequently, the optimization of the entire technology shall be obtained by a tuning work among the various parameter values or choices of the operations of a technology. The intranality effect also occurs in the change of an operation in the structure of a technology that may influence the efficiency of other operations of the structure. In this case, the operative conditions of other operations of the structure shall be changed in order to introduce modifications of the structure of the technology that is necessary for an efficient use of the innovation. This effect can be easily faced when all the technological operations are made in the same firm, but the situation is different when certain operations are subcontracted to other firms, as often occurs in industrial districts producing the same product. In this case, an innovation of a producing firm might necessitate changes in the operations of a subcontracting firm in order to use the innovation. For example, such a change might be unwanted by the subcontracting firm because it necessitates investments or because it would influence negatively subcontracted work for other firms. An example of intranality effects combined with externality effects was observed with the appearance of a new type of brass composition complying with new environmental regulations for the production of valves and faucets, the history of which have been reported as follows:

> In the 90' in USA and in other countries were introduced strict norms about contamination of drinking water by heavy metals, in particular lead. Valves and faucets are in fact made using a brass containing lead in order to improve the machining speed, but normal content of lead would contaminate water in certain cases above the limits of the norms. Solutions were the use of a treatment able to eliminate the lead existing on the surface of brass, or to develop a new lead free, easy machining, brass alloy. Such last solution was developed by an important German producer of brass with an alloy called Ecobrass®. Unfortunately, such alloy contained silicon

giving problems to the chroming operation used for faucets, operation normally subcontracted, that would necessitate a specific bath treatment to eliminate silicon from the surface. However, such additional treatment was expensive and the bath was difficult to handle because very aggressive and then not accepted by subcontractors. In this situation only producer of valves that do not carry out any chroming operation were interested in the use of Ecobrass®. Actually, although it exists a certain use of Ecobrass®, because of the cost of this alloy, many producers of valves tried to modify their machining operation in order to obtain acceptable speeds with simple free lead brass, or used an additional operation consisting in a special treatment to eliminate the lead on the surface of the brass.

Another example of intranality effect has been observed in the study of the innovation processes of an Italian industrial district of production of ceramic tiles in which a new product or process developed by a firm has been observed, necessitating complementary innovations in subcontracting firms that would be adoptable only if an important demand were generated for the firms that should introduce the complementary innovations [19].

2.5.3 Ramification of Technologies

This important technological process occurs when a new technology with an important radical degree appears in the space of technologies and triggers the formation of other technologies that represent improvements, diversifications, or alternatives to the initial radical technology. All of these technologies, represented by points in the space of technologies, may be connected by forming a ramified structure evolving with time. It should be noted that ramification of technologies is not a process of diffusion of technologies in which technologies are not changed beside minor variation in operational parameters, due to different environment in which the technology operates. Technology ramification is normally characterized by a decrease in the radical degree of the formed technologies as far as they are distant from the original technology, and by an increase in their number. A schematic example of ramification of technologies is reported in Figure 2.6. In this figure, the traits of ramification connecting two points (technologies) represent modifications of the structure of a previous technology that changes the Hamming distance in the space. A new technology may also contain operations not existing in the previous linked technology, but from another technology of the space that is represented by a trait connecting two branches of the development of the initial technology. It should be noted that such connections between branches, compared with biological ramifications, do not exist in natural trees as in biological evolution, in which it would correspond to a combination of genes of different species that is not allowed, but is in fact possible using biotechnologies. Such branching differences between evolution of technology and that of biology have also been noted discussing the factors of selection among technologies [20]. An indirect demonstration of existence of technology ramifications may be found by studying the formation of patents from an initial patent of a radical innovation, and connecting it with new patents covering improvements, diversification, or alternatives to previous patents. This study has been conducted, for example, to considering the topological evolution of patents from an initial patent covering computerized tomography from 1975 to 2005 [21], showing the important development of ramifications, and including the existence of connections between branches of the ramifications. Another indirect representation of ramified evolution of technologies may be found in studying the formation of start-ups by spin-off from an important initial technological company. That has been observed in the case of Fairchild Semiconductors, an important firm of the Silicon Valley in the field of microelectronics. In fact, from 1959 to 1971, it was possible to observe the direct or indirect formation of 35 start-ups, while the total number of start-ups that may be traced until 2014 is 92 [22]. The formation of ramifications and branch connections is not limited to technologies that fulfil the same purpose but are possible lateral ramifications in other spaces for technologies that in fact have different purposes. We have previously cited the example of the coffee maker Moka Express® derived from a technology of laundry washing. Another important example is the case of additive manufacturing; this technology of making objects was originated by ink jet printing technology. Initial additive manufacturing was at first operating at room temperature with materials melting at relatively low temperature as plastics, but the technology was developed further for medium- and high-temperature operations making additive manufacturing also possible with ceramics and metallic materials. In addition, printing technology developed further with laser printing. Another important observation that may be made on ramification concerns the origins of the various involved technologies. The triggering technology is often an application of results of scientific research. New technologies obtained in the proximity of the triggering technology are often the result of R&D activities. The appearance of new technologies in great numbers is the result of innovations of a more incremental nature, often obtained by LbyD during the use of the technology. In this case, the inventive activity is not necessarily linked to researchers, but rather to technicians operating the technology. The more external part of the ramification consists of a large number of technologies that, although with limited applications, in fact represent the bulk of the socio-economic impact of the technology, showing the full development of the initial radical technology. Consequently, because of the technical origin of these last technologies, intermediate scientific and technical education becomes of great importance to assure the prosperity of a country based on technology innovation. This has been demonstrated by studies that have shown that education represents a factor for economic growth that is more important than industrial property or profit growth [23].

2.5.4 The Spandrel Effect

The spandrel effect explains the formation of new radical technologies, not necessarily with the same purposes, in the space of technologies. Their further

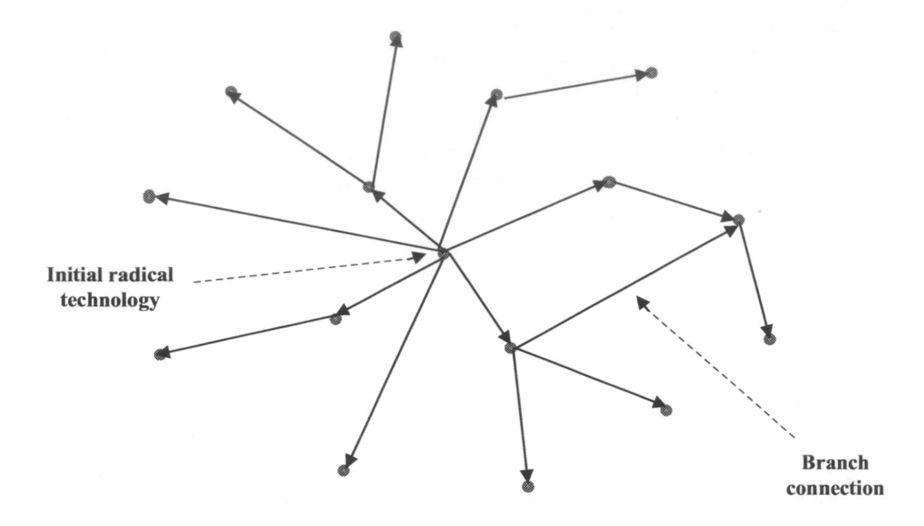

FIGURE 2.6 Schematic view of ramification of technologies.

ramification thereby impacts the evolution of technologies. This effect has been originally recognized in the biological evolution by Stephen Gould, an evolutionary biologist, who used the concept of spandrel to explain the existence of an alternative means of Darwinian selection in the evolution of species by formation of living organisms as a consequence of other features not directly linked to the Darwinian selection process, in fact occupying free niches in the space of biological evolution. The spandrel is an architectural space that is formed fatally erecting for example a dome on a square flat roof, and is represented by the four free spaces remaining at the four angles of the roof. Such spaces may be used by architects to erect statues and ornamental parts in order to improve the aesthetics of the dome. It is interesting to know that Gould got the idea of using the spandrel concept for evolutionary biology by observing the San Marco Cathedral domes in Venice, where spandrel spaces are used as further ornamental elements of the architecture [24]. The spandrel concept may be used analogously for technology evolution by considering the ramification process of technologies as analogous to a dome space, and roof space as an extended space of technologies also including operations that are not present in the technologies of the ramification, but may be used for the development of radical technologies with the same or a different purpose. The space of technologies, outside the space occupied by the ramification, may be considered the *spandrel* of the extended space of technologies, and it is in this space that new radical technologies might appear to have the same purpose, or different purposes as in the case of additive manufacturing originated by ink jet printing cited previously.

2.5.5 Velocity of Innovation

The velocity of innovation is represented by the number of new technologies that appear within a certain time in a technological sector for various applications. This velocity depends on a great number of factors concerning economic situations, such as chosen strategies, availability of private capitals of industrial or venture nature, public aids, and high innovative potential in certain sectors. In certain cases, however, such as in development of new drugs or medical apparatus, the velocity of innovation may be limited by protocols of testing procedures with long time periods necessary to obtain the commercialization permit for the product. Among the various factors determining the velocity of innovation, there are also some of a technological nature. In fact, the model of technology considers an innovation as a change to the structure of a technology and the innovation velocity also depends, from a technological point of view, on the velocity with which it is possible to change the operations during the development of a new technology. Such velocity of change is dependent on the physical, chemical, or biological nature of the operations. It is easy to understand that the change in an informatic program or in assembling new components of an electronic circuit may be more rapid than changing equipment in a pilot plant of a new chemical process, or machining a part for a new mechanical artefact. In the case of biotechnologies, the velocity of innovation is limited by typical long periods of biological processes that cannot be necessarily accelerated by catalysts or by elevating the temperature as in chemical processes. In general, the velocities of innovation are not a subject of studies for the complexity of influencing factors; however, it is possible that the enormous development of ICT is also due to the rapidity of changes in software and hardware of ICT devices with respect to the innovation velocities observed in the case of development of nanotechnologies or biotechnologies that include construction of specific equipment and are perhaps characterized by slow biological processes. In fact, there is an observed important effect of innovation velocity in the field of ICT that influences the implementation of these technologies during an implementation project. The high rates of appearances of new technologies in this field, might make the technologies chosen for the implementation project obsolete. This type of situation was observed at the end of the 1990s in firms offering internet services to client companies for commercial or advertising purposes. In particular, the case of the Silicon Alley, an area of New York with a high concentration of firms offering ICT services, has been studied,. An anthropological study, had the aim of finding which solutions were elaborated by the firms to face the effect of high rates of innovation described previously, and to find which solution was the best reorganization of project management, resulting in a Darwinian selection of the various solutions applied by the firms [25]. The optimal organization that was found for the projects was called *heterarchical*, in opposition to the hierarchical organization used in conventional project management. This new organization was characterized by the availability of highly diversified competences, interdependence, and a certain heterogeneity of the organization. Such an impact of the velocity of innovation on the implementation projects related to ICT might also occur in other fields such as finance, administration, and even in implementation of enabling industrial technologies linked to ICT.

2.5.6 The Red Queen Regime

The Red Queen regime is a situation often observed in industrial districts or sectors making the same type of products and it is related to competition among firms in the field of technology innovations. Red Queen regime is a term used originally in description of genetic competition between preys and predators [26] and the Red Queen is a character of Lewis Carroll's *Through the Looking Glass*," the continuation of *Alice's Adventures in Wonderland*, in which Alice is told, "In this place it takes all the running you can do, to keep in the same place." The competitiveness of firms depends on many factors such as strategies, efficient productions, and marketing. However in certain cases, technological aspects may become predominant for a firm to succeed. The model of technology may explain certain aspects of the origin of technological competitivity by considering the operations of the technology structure, the degree of radicality of an innovation, the fact that operations are associated with competences that are necessary for efficient use of the technology. These competencies during a technological competition may be more or less available or may take time to be acquired. Taking in consideration an industrial district or sector making similar products, all firms own approximately the same competences because using similar technologies. If one of these firms introduce improvements or incremental innovations of the technology in term of process or product, it obtains a technological advantage in respect to the other firms. However, because of the low degree of radicality of the innovations, the necessary competences are also owned by the other firms or quickly obtained. The other firms may then easily react and introduce similar or alternative innovations and possible existence of patents protections might be easily bypassed because of incremental nature and low level of novelty of the patents. The result is that the initial acquired technological competitive advantage of the firm disappears and, with time, the repetition of such situations leads to a continuous technological development of incremental nature of the district or of the sector without any important economic growth or appearing of a technological leader firm. Such situation of technological nature is called Red Queen regime. On the contrary, if a firm of the district or sector, develops a new technology with a high degree of radicality, and then including operations quite different from those of the pre-existent technology, the other firms will have a lack of competences dues to the difference existing with the operations of the new technology. These firms would have many difficulties to counter the innovation, possibly also because of a strong position of patents with a high level of novelty, and consequently the necessity to make strong efforts to become anew technologically competitive. The Red Queen regime is in fact a dangerous regime because of the possibility of disruptions by the appearance of a radical innovation that makes obsolete previously used technologies putting in danger the existence of firms that cannot counter the new situation. A historical example of rupture of a Red Queen regime occurred in the 1970s when the Swiss watch production district was threatened by the Japanese watch industry; the detailed history is reported next:

> Swiss watch industry was composed in the seventies by a great number of small companies, organized as an industrial district in the north west of the country, and using mechanical technologies for watches production. Innovations were essentially incremental and, although the use of quartz piezoelectricity was known, it was applied only to a few types of luxury models as considered expensive. Japanese watches industry oriented technical developments in a radical direction using quartz piezoelectric oscillations instead of traditional mechanisms, a digital indication of hours using liquid crystals, a material that change its luminosity as a function of applied voltage, and introducing a small battery supplying energy to the watch. This product had a relatively low price and reached a great success in the market putting in great difficulties the traditional Swiss watch industry and, at the end of the seventies, about 40% of Swiss watch firms disappeared. Survival and restarting of Swiss watch industry may be attributed essentially to the action of Nicholas Hayek that organized the merging of many watch firms in the SMH holding, and developing a new concept of watch as ornament, the Swatch®, based technologically on a cheap quartz system with an industrialization that lasted about four years. Swiss watch industry did not have any liquid crystal technology and practically never used digital indications of hours in its models.

It may be observed that the recovery of the Swiss watch industry may also be attributed to the fact that a certain knowledge of quartz technology existed in Swiss research laboratories, which made a relatively rapid beginning of the industrial production of Swatch® possible; otherwise, the recovery of the Swiss watch industry would be much more difficult.

2.5.7 Transfer of Technology and Know-how

Technology transfer in fact commonly refers to two different technological processes, the first one concerns bringing a new technology into use after its development, the second one concerns the transfer of a technology used in an industrial plant to another plant in another location, or in a more general way, the transfer of technology of use of a process or an artefact from an expert to a newcomer. These two processes present similarities as well as differences. In both processes there are, in most cases, some changes in the transferred technology to make it suitable for the new operative conditions. These changes are not necessarily only at the level of values of parameters to be used, but possibly also consist of the introduction of limited innovations of an incremental nature; they are normally caused by the differences in externalities that exist with the new application or in the environment of the new location of the transferred technology. For the description of the process of technology transfer it is previously necessary to detail the relation between a technology and its operations; it is important to consider an aspect that is necessarily associated with the transfer of a technology constituted by the *know-how*. This concept may be well defined and described using the model of technology. The know-how represents a specific knowledge acquired by personnel operating a technology, particularly during the use of technology and LbyD activities. The know-how is of course different in the case of technologies producing or using an artefact. In order to describe the knowhow using the model of technology, we have to consider a technology in action operated following an optimal recipe corresponding to a *peak* in its technological landscape. This technology is normally influenced by externalities, in the major part of cases with limited effects, which nevertheless change the technological landscape to some extent. Operators in most cases simply changes the parametric values in order to restore optimal conditions using their technological or scientific knowledge, as well as accumulated experience. As the effects of externality may be of different types and appear and disappear many times, the necessary changes to maintain optimal conditions of operation for the technology are memorized by operators and constitute their know-how of the technology. An important aspect of know-how is that it is composed by a great number of elements of acquired knowledge that makes their complete transfer in spoken or written form to another operator who begins the use of the technology practically impossible. For these reasons, economists who study the economy of innovation consider know-how a tacit knowledge. The transfer of knowhow associated with a technology, either during entering into use or during transfer of location, is facilitated by manuals and oral instructions, but is essentially based on imitation and on the direct experience of the operator, who is willing to acquire the knowhow. This represents an essential element to succeed in technology transfer in the operation of, for example, devices, instruments, production plants, and factories. In fact, for the technology transfer of already used technologies, it is necessary to consider the training of personnel who shall operate the technology, not simply through manuals and oral instructions, but through imitation and direct experience normally carried out for some time with the aid of personnel who already have knowhow of the transferred technology. In the case of technology transfer brought about by bringing a new technology into use, it is necessary to find solutions for the various conditions of the transfer, such as the increase of production capacities and different externalities in which the technology operates with respect to the operations carried out at the level of the laboratory or pilot plant, or through fabrication of prototypes. This means that the knowhow acquired during the development of the technology is generally not sufficient to operate the technology at an industrial level and it shall be acquired through a LbyD activity with a sequence of improvements that may be expressed by the so-called learning curve. The problems associated with transfer of technology are important and it is not rare that new technologies entering into use are abandoned because of the lack of possibility of reaching satisfactory operative conditions at the level of an industrial demonstration plant.

2.5.8 Model of Technology and Patents

Patents are a typical topic discussed in technology management, but the model of technology may suggest some interesting aspects of patents that may be useful in the understanding of their relation to the innovation process. A patent represents a legal right, described with a technical language, granted by an administration, establishing a real right with respect to an invention. An invention represents a technological innovation that allows the improvement of existing technologies, creating new practices. It is distinguished by a discovery that is a typical result of scientific activities, revealing pre-existing things or enriching the general knowledge of scientific nature. It should be noted that the given definitions and scopes covering patents, although generally accepted, might be the object of some different interpretations following the administrations of various countries; this may occur, for example, in the field of definition of invention of new matter, software inventions, results of artificial intelligence, inventions of a biological nature, particularly those concerning biotechnologies and applications of synthetic biology. Considering the definition of technology of the model as a structure of operations associated with instructions for the various parameters, the scope of a patent covering a new technology may be seen as a delimited space in the corresponding technological space and landscape, in which are included the instructions characterizing the invention and reported in the examples and in the claims of the patent. It should be noted that the efficiency of recipes of the patented technology is not generally taken into consideration in a detailed and quantitative way. In a patent, economic aspects and technical improvements are normally given in a very generic way. This is the result of the aim of the writer of the patent to give minimal information about the real efficient way to operate the technology, but nevertheless also covering possible improvements or unwanted imitations of the technology by other inventors. These aspects are also important because patents are often applied after a feasibility study before full development studies make the indication of the best operative conditions of the technology possible. In conclusion, it should be noted that study of patents from the technological point of view is not necessarily limited to verifying the patentability of an innovation but, following the model of technology, may consist in patent intelligence studies in which exploration of the technological landscape covered by a patent might be of interest for suggesting different conditions or innovations that are not covered by the patent.

2.6 THE STAGES OF THE INNOVATION PROCESS

After the discussions about the various technological processes, it is possible to present another view of technology innovation that describes the sequence of the various stages of development existing from the generation of the innovative idea to the use of the new technology. A view of technology innovation as a process appeared for the first time in the 1960s [27]. In this case, the innovation process was explained as a sequence of three phases constituted by basic (oriented) research, applied (industrial) research, and industrial development, also seen as precompetitive development. This view of the innovation process has been accepted as a standard reference for surveys on research and experimental development in an OECD guide with the title *Frascati Manual*. This manual appeared in various editions, the last of which was in 2015 [28], this definition is currently used for socio-economic studies on research and innovation. In fact, the *Frascati Manual* considers the activities of innovation as R&D activities defined as creative and systematic works undertaken in order to increase knowledge, including knowledge of humankind, culture, and society, to devise new applications of available knowledge. This defined activity is of three types: basic research, applied research, and experimental development. However, such activities are not considered in a temporal sequence because of the existence of various directions of flows of knowledge. Basic research is experimental or theoretical work that is undertaken primarily to acquire new knowledge without any particular application in view, and may be distinguished from oriented basic research, which is carried out with the expectation that it will produce knowledge useful for solutions of future problems, as well as pure basic research, which is carried out for the advancement of knowledge, without seeking economic or social benefits. Applied research is original investigation directed towards a specific objective for possible applications to products and operations. Experimental development is systematic work, drawing on knowledge gained from research and practical experience, producing additional knowledge, which is directed to producing new products or processes. In technology dynamics, the view of the innovation process is more structured and constituted by a sequence of steps starting from the generation of innovative ideas to the use of technology, also considered a source of new technologies. The innovation activities considered in the *Frascati Manual* corresponds, in a certain way, only to the intermediate stages of the innovation process described by technology dynamics. It is important to observe that, differently from in the *Frascati Manual*, scientific research is not integrated into the innovation process considered by technology dynamics, but constitutes an interface of the process supplying knowledge and research results for generations of innovative ideas accordingly to the nature of generation of new technologies [2]. In fact, the OECD model is not considered of interest in discussing R&D activities with the objective of development of new technologies in R&D management [29]. Actually, in technology dynamics innovation is presented as a process that transforms an innovative idea in a new technology through a temporal flux of activities divisible in various steps. Following technology dynamics, the innovation process may be divided into five steps: generation of the innovative idea, feasibility study, development study, industrialization, and use of the developed technology. These stages are represented schematically in Figure 2.7. Such stages are characteristic of the R&D activity, but they may be also considered for innovation processes not exploiting scientific results for the generation of the innovative idea, but only for new combinations of pre-existing technologies. In conclusion, a technology

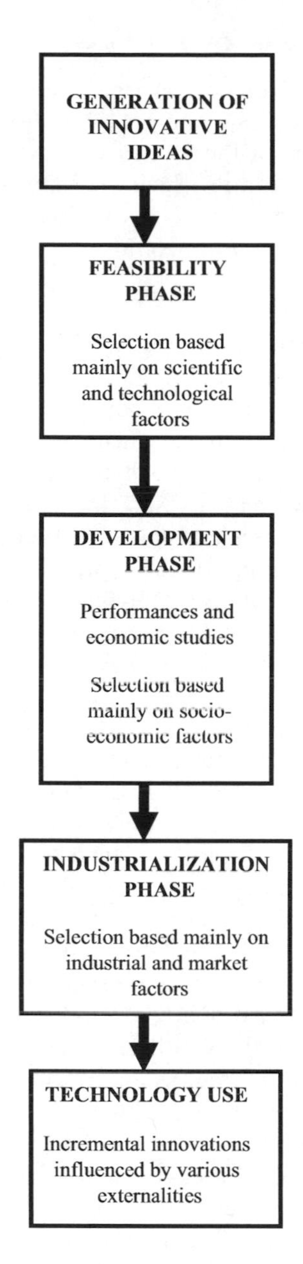

FIGURE 2.7 Steps of the technology innovation process.

may be considered as a likely biological entity that is born by an innovative idea, developed until the stage when it can enter into use, continued with modifications induced by externality factors, finally becoming obsolete and possibly be substituted by a new more efficient technology.

2.6.1 The Generation of Innovative Ideas

There are two aspects of the process of generation of innovative ideas besides its combinatory nature with a possible exploitation of scientific results: the first one concerns the actors of the inventive idea that may be a single person or several persons in a situation in which the innovative idea emerges through generative relations, the second one concerns the climate of the environment that favours the generation of innovative ideas.

2.6.1.1 Individual generation

The individual generation of an innovative idea was the common and nearly unique way to make inventions in the past. Typically, the innovation potential of an individual person depends on the individual's creativity, which has been studied particularly in the context of management of R&D, especially for industrial R&D laboratories. In fact, Dumbleton's handbook on R&D management [29] dedicates an entire chapter to creativity, showing how individual creativity is composed of various complex phases and favoured by a suitable environment with trust, free flux of information, and possible autonomous decisions by personnel. An interesting original contribution to creativity in inventions may be found in the studies of Genrich Altshuller, a Russian engineer who studied a huge number of patents for many years, making detailed classifications of their various inventive characteristics. He developed a method called TRIZ, a Russian acronym, meaning the science of the theory of invention [30]. This method is normally considered a typical tool used in management of innovation, as for example, the OFD or Taguchi methods. However it is not limited to quality of products and their markets but is also used for the development of innovative products, their protection, and anticipation of future products. What is of interest in the TRIZ method, in relation to technology dynamics, is the definition given by Altshuller to the *inventive problem* considered as a problem containing at least one contradiction, i.e., a situation in which a tentative improvement of a characteristic of a system brings about the worsening of another characteristic corresponding to the effects of intranality described in the model of technology. Furthermore, Altshuller, in the definition of the various levels of innovations and their statistical distribution, found that innovations linked to new scientific discoveries were very limited, and that most innovations in fact consisted of improvements or alternatives of existing practices and even innovations coming from knowledge existing in other technologies, all that coherent with observations of technology dynamics with respect to the generation of new technologies and their ramifications. Finally, it should be noted that the relations among inventors of new technologies, based on scientific discoveries, scientific community opinion may be quite complicated as sometimes a development based only on trial-and-error may lead to important innovations against the opinion of scientists based on available scientific knowledge. In fact, this was the case of Guglielmo Marconi regarding the oceanic transmission of electromagnetic waves, described by George Basalla in his book on the evolution of technology [20], which merits to be described in some detail:

> Guglielmo Marconi, after having shown the transmission of electromagnetic waves, was interested to show the possibility to have transmission at very long distance, in particular on oceanic distances allowing communication of signals among ships and coastal stations constituting an enormous field of applications of its invention. Actually, the physics of Hertzian waves would not allow transmission beyond the horizon. That might be overcome on earth by using repeater antennas but that it is impossible on the sea. Marconi, that was not an academic figure although not deprived of a scientific knowledge, tried nevertheless an oceanic transmission from Poldhu in Cornwall and Clifden in Co. Galway in Ireland, and he was able to announce that the message was received at Signal Hill in St John's, Newfoundland on December 12, 1901 opening the great field of applications of wireless communications. Of course, scientists were not in error about limits of horizon, but they did not really know what happens to waves in the open space and did not know the existence of an ionized stratospheric layer reflecting the waves.

Actually, Guglielmo Marconi was not a scientist but an inventor and entrepreneur who decided not to apply scientific knowledge to the solution of a technical problem and provide technological solutions for problems not yet fully comprehended by the scientific community [20].

2.6.1.2 Generative relations

Inventions emerging by generative relations among actors interested in making innovations have become more and more important as may be observed by the presence of a high number of patents presenting more than one inventor. Generative relations leading to inventive ideas are formed by two or more representatives, for example, between a technical representative and an industrial or commercial interlocutor discussing innovation potentialities or problem solving, and also among two or more researchers discussing potential applications of knowledge coming from scientific research or R&D activities. The emerging of an innovative idea during a generative relation has been studied in the case of ROLM, a small company of the Silicon Valley, which decided in the 1970s to diversify its activity toward technologies of introduction of computer capabilities in telephonic systems used by firms [31]. Many innovations introduced in this technology emerged from discussions between technical and commercial representatives of ROLM and telecommunication managers of client firms. The generation process was studied for this particular case, developing a specific general model for the generative relation [31]. The model defines a certain number of elements and how these elements interact in the generation of innovations indicated as follows:

Agent: any individual or group of individuals that interact in a system constituted for example by firms, departments, clients, researchers, and interviewers.

Artefact: any product, process, or service that may be the object of exchange among the agents, including projects, financial instruments, and means of communication.

Attribution: any interpretation that an agent has of himself, of the other agents, and of the artefacts.

Generative relation: any relation among the agents or also among artefacts that are able to induce changes among the parts concerning the attributions of agents or artefacts in such a way that new entities or innovations are created.

The generative process is essentially the following: at the beginning of the relation, the agents, who do not necessary belong to the same organization, have quite different ideas on artefacts and their attributes. During generative relations, the agents change their ideas and converge toward an agreement on the innovation. In the generative relation, it is important that in the beginning there is a concrete subject of discussion, and not simply a presentation of, for example, experience, products, equipment. without a specific significant objective. The concrete initial subject, during the generative relation, might be completely changed, converging toward a completely different subject that leads to a different type of innovation. This model is in good agreement with the experience of generative relations that exist, for example, during discussions between researchers and industrial partners about a proposal for an R&D project, or among researchers discussing a possible research proposal to be submitted for financing.

2.6.1.3 Environment for generation of innovative ideas

We have already seen that individual generation of innovative idea is promoted by a favorable environment existing in a research organization characterized by free flux of information and autonomy of the researchers. A favorable environment for the promotion of generation of innovations may exist also at the level of a territory, and an example has been given in the book of Annalee Saxenian describing the innovative culture existing in the Silicon Valley that competed successfully with Route 128 in the Boston area [32]. Silicon Valley is described as an inventive technical community that is not limited by the hierarchical structures existing in competing firms of Route 128. Another important observation that can be made for promotion of innovative ideas is the fact that creativity and generative relations are accompanied and favoured by an entrepreneurial mentality, existing not only in terms of individuals' initiatives for the development of their own innovative ideas or creation of start-ups, but also for employed people attentively looking for innovations in their activity. This aspect was well understood, for example, by Clyde Williams, director of the Battelle Columbus Laboratories in the 1930s, who created and promoted the figure of the *researcher entrepreneur* who was able to generate new innovative ideas from his work in R&D projects [4]. The promotion of entrepreneurship in people for the generation of innovative ideas does not depend particularly on availability of financial supports, but rather on educational experiences, especially among young people, through the favorable environment in which they live or through stages and study tours in organizations or territories in which the association of creativity and entrepreneurship is well developed as, for example, in the Silicon Valley. We present as example the activity and results of an Italian association "La Storia nel Futuro" that organized study tours in the Silicon Valley for students, and the obtained results in terms of generation of start-ups by participants to the tours. A brief history of the association is reported below:

> The association "La Storia nel Futuro" was founded in 1998 by Paolo Marenco, at that time director of a technological parc in a territory at the north east of the Italian Piedmont region, and Lino Cerutti, a journalist specialized in the history of this territory. The first activities were the organization of conferences with speakers active in the territory or emigrated abroad with the aim to promote culture and entrepreneurship by activation of the "genius loci" of the territory through description of present or past experiences. These conferences were later extended to the entire Italian territory. The beginning of the activity of study tours in the Silicon Valley was the result of a casual meeting in 2004 of Paolo Marenco with Jeff Capaccio, an Italian-American lawyer, secretary of the just constituted SVIEC, the Silicon Valley Italian Executive Council, now having more than 1500 members. During the meeting it was born the idea to use the SVIEC for the organization of study tours in the Silicon Valley, with the participation of Italian students terminating their university studies. The tours consisting in visiting firms and universities of the Silicon Valley and meeting Italian executives, managers and researchers working in this territory. There was then the possibility to have discussions, mainly in Italian language in order to have a full understanding about the organization and working climate existing in the Silicon Valley. It was then expected in this way a promotion of creativity and entrepreneurship in the students, and possibly even the creation of start-ups by participating students entering in the world of labor. From 2011 these study tours have been extended to Italian managers, entrepreneurs and start-up founders. In the last years the study tours had a certain international extension in participations, and they have been also extended to all university and high school students believing in the importance to make available this experience also to very young people.

The results of study tours of this association have been the object of a case study about promotion of entrepreneurship in the new technologies during the period of activity between 2005 and 2017 [33]. In this period, 16 study tours were organized for the participation of 314 students, and 15 study tours for a total of 265 managers and entrepreneurs. The formation of 20 start-ups have been observed in which the founders confirmed the importance of participation in the tour in their decision to realize the start-up. In Figure 2.8 we report the partition of types of education of the participating students, and in Figure 2.9 the partition of the types of start-ups generated. The ratio between the number of generated start-ups and the number of participating students is around 7%, a value that may be considered satisfactory. It may be observed by comparing results of Figures. 2.8 and 2.9 that a quite large partition of different types of studies led to equally diversified partition in the type of formed start-ups, showing that effects of study tours in generation of start-ups have similar effects for all types of student education.

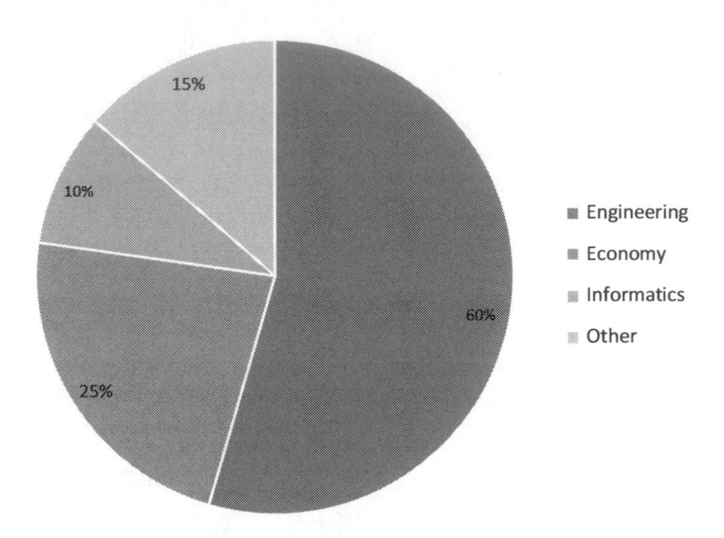

FIGURE 2.8 Partition of types of education of participating students.

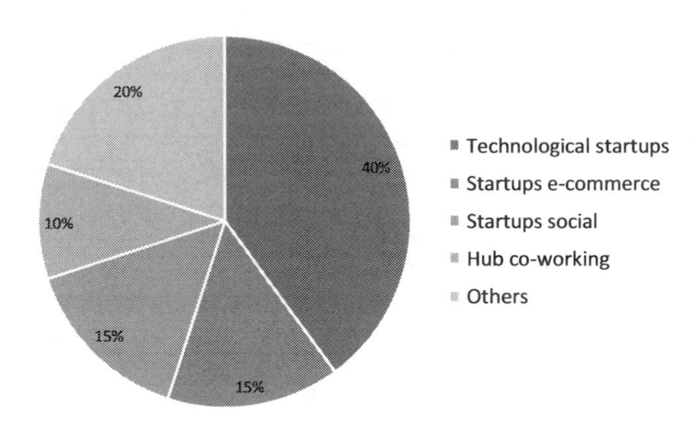

FIGURE 2.9 Partition of types of generated start-ups.

2.6.2 Feasibility, Development, and Industrialization

The stages of feasibility, development, and industrialization represent the core of the innovation process leading to the formation of new technologies.

2.6.2.1 Feasibility studies

This step concerns the feasibility of the innovative idea for the considered application, often resulting from scientific research and constituting the typical initial step of the R&D activity. The scientific and technological aspects are of major importance in determining the continuation or termination of the innovation process.

2.6.2.2 Development studies

This step principally concerns the determination of the performance level of the innovation, with respect to required specifications, as well as the evaluation of its economy. This step may include operations with pilot plants or construction of prototypes and many innovations with a pure combinatory character or of incremental type may start developments at this stage. Socio-economic factors are of major importance in respect to technical factors in the decision to continue or to stop development. This stage is considered the most difficult to overcome for the innovation and it is sometimes called *The Valley of Death* of the innovation projects [34].

2.6.2.3 Industrialization

This stage includes the final development of the new technology and the planning for its industrialization; in particular, it may also include the planning of construction of demonstration plants in order to show the validity of the new technology.

2.6.3 Use of the Technology

With the industrialization stage, the development process of a new technology is normally considered terminated, but not necessarily the innovation process. In fact, the model of technology considers the possibility of generation of innovations during the use of a technology, particularly during the activity of LbyD, under the influence of externalities that stimulate, not only the search of optimal conditions in the technological landscape, but also typically incremental innovations, as reported previously, considering the effects of externalities on the operation of a technology. The use of a technology is also a source of innovative ideas concerning diversifications and alternatives of the technology constituting the base of the ramification process of technologies; it may even be the source of formation of lateral ramifications concerning technologies with a different purpose, as for example, the cited technology of additive manufacturing generated by ink printing technologies.

2.7 RISK AND UNCERTAINTY OF THE INNOVATION PROCESS

An important aspect of the innovation process is represented by the risk of failure, and this risk changes during the various stages of the innovation process. When discussing innovation processes, it is important to distinguish the concept of risk from that of uncertainty [35]. In fact, uncertainty represents the impossibility of estimating risk in terms of its probability of success or failure.

Actually, the progress of an innovation process transforms uncertainty into risk, making possible decisions about continuation or termination of the innovation development. There are various types of uncertainties or risks in the various stages of the innovation process that concern technical aspects, performance, economy and market [36]. Technical uncertainty is reduced mainly in feasibility studies, uncertainty of performance is reduced during the development study and before the economic uncertainty, market uncertainty is the most difficult to reduce and it occurs principally during the use of the technology. Finally, we report indicatively in Figure 2.10 the typical reduction occurring with time of an initial number of innovation projects typically concerning the R&D activity. Technology developments are abandoned at different rates following the feasibility, development, and industrialization steps, reaching a maximum reduction rate for the development step, the cited *Valley of Death* of R&D projects. It should be noted that the development step also has the longest duration with respect to the other steps. Finally, it may be observed that in fact, only a very minor number of R&D projects reach the industrialization step.

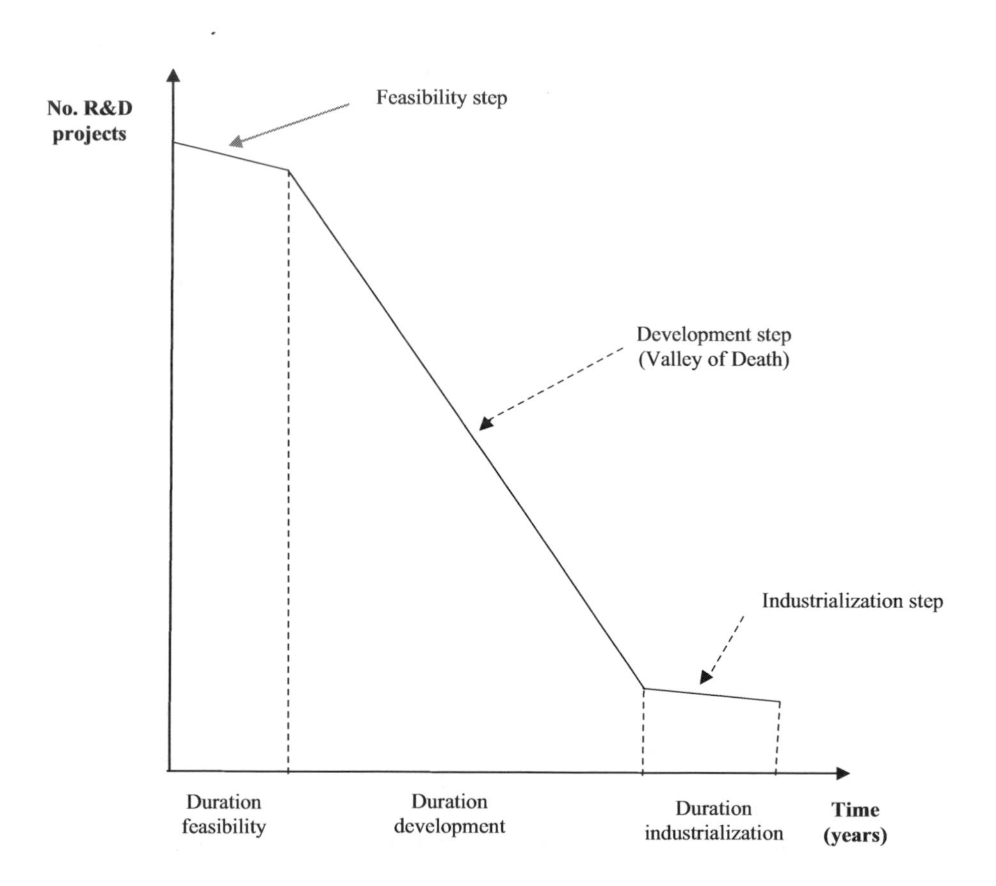

FIGURE 2.10 Indicative evolution with time of an initial group of R&D projects.

REFERENCES

[1] Evans D. 2003, Aristotle on *techne*, 37–47, in *Les philosophes et la technique*, Chabot P., Hottois G., Editors, Librairie Philosophique J. Vrin, Paris.

[2] Arthur B. 2009, *The Nature of Technology*, Free Press, Division of Simon & Schuster Inc. New York.

[3] Bonomi A., Marchisio M. 2016, Technology modelling and technology innovation: how a technology model may be useful in studying the innovation process, *IRCrES Working Paper*, 3/2016.

[4] Boehm G., Groner A. 1972, *Science in the Service of Mankind, the Battelle Story*, Lexington Books, D.C. Heath and Company, Lexington, DC.C.

[5] Isaacson W. 2011, *Steve Jobs*, Simon & Schuster New York.

[6] Hall B.H., Lotti F., Mairesse J. 2009, Innovation and productivity in SMEs: empirical evidence for Italy, *Small Business Economy*, 33, 13–33.

[7] Waldrop M. 1992, *Complexity, the emerging science at the edge of order and chaos*, Simon & Schuster, New York.

[8] Auerswald P., Kauffman S., Lobo J., Shell K. 2000, The production recipe approach to modeling technology innovation: an application to learning by doing, *Journal of Economic Dynamics and Control*, 24, 389–450.

[9] Kauffman S., Levin S. 1987, Toward a general theory of adaptive walks on rugged landscapes, *Journal of Theoretical Biology*, 128, 11–45.

[10] Frenken K. 2001, *Understanding Product Innovation Using Complex Systems Theory*, Academic Thesis, University of Amsterdam jointly Université Pierre Mendès France, Grenoble.

[11] Wright T.P. 1936, Factors affecting the cost of airplanes, *Journal of the Aeronautical Science*, 2, 122–128.

[12] Kauffman S., Lobo J., Macready G.W. 2000, Optimal search on a technology landscape, *Journal of Economic Behaviour and Organization*, 43, 141–166.

[13] Lobo J., Macready G.W. 1999, Landscapes: a natural extension of search theory, *Santa Fe Institute Working Paper*, 99-05-037.

[14] Fleming L., Sorenson O. 2001, Technology as a complex adaptive system: evidence from patent data, *Research Policy*, 30, 1019–1039.

[15] Strumsky D., Lobo J. 2002, “If it isn’t broken, don’t fix it”: extremal search on a technology landscape, *Santa Fe Institute Working Paper*, 03-02-003.

[16] Fleming L., Sorenson O. 2004, Science as a map in technological search, *Strategic Management Journal*, 25, 909–928.

[17] Nelson R., Winter S. 1977, In search of a useful theory of innovation, *Research Policy*, 6, 1, 36–76.

[18] Dosi G. 1982, Technological paradigms and technological trajectories. A suggested interpretation of the determinants and direction of technical change, *Research Policy*, 11, 147–162.

[19] Russo M. 2003, *Innovation Processes in Industrial Districts*, ISCOM Project, Venice November 8–10, 2002 (revised version 07.02.2003).

[20] Basalla G. 1988, *The Evolution of Technology*, Cambridge University Press. Cambridge UK.

[21] Valverde S., Solé R., Bedau M., Packard N. 2006, Topology and evolution of technology innovation networks, *Santa Fe Institute Working Paper*, 06-12-054.

[22] Morris R. 2014, The first trillion-dollar startup, *Endeavour Insight Monthly Newsletter*, Jul. 26, 2014.

[23] Wang E.C. 2010, Determinants of R&D investment: the extreme bound-analysis approach applied to 26 OECD countries, *Research Policy*, 39, 103–116.

[24] Gould S.J. 1996, The pattern of life’s history, in *The Third Culture*, Brockman J., Editor, 51–73, Simon & Schuster, New York.

[25] Girard M., Stark D. 2001, Distributing intelligence and organizing diversity in new media projects, *Santa Fe Institute Working Paper*, 01-12-082.
[26] Van Valen L. 1973, A new evolutionary law, *Evolutionary Theory*, 1, 1–30.
[27] Freeman C. 1974, *The Economics of Industrial Innovation*, Penguin, Harmondsworth, OECD Report: The Measurement of Scientific and Technical Activity.
[28] OECD. 2015, *Frascati Manual 2015: Guidelines for Collecting and Reporting Data on Research and Experimental Development*, The Measurement of Scientific, Technological and Innovation Activities, OECD Publishing, Paris.
[29] Dumbleton J.H. 1986, *Management of High Technology Research and Development*, Elsevier Science Publisher, New York.
[30] Altshuller G.S. 1988, *Creativity as an exact science*, Gordon & Breach, New York.
[31] Lane D., Maxfield R. 1995, Foresight, complexity and strategy, *Santa Fe Institute Working Paper*, 95-12-106.
[32] Saxenian A. 1994, *Regional advantage*, Harvard University Press, Cambridge.
[33] Bonomi A. 2018, Promozione dell'Imprenditorialità nelle Nuove Tecnologie, Caso Studio: Associazione La Storia nel Futuro *IRCrES Working Paper*, 9/2018.
[34] Auerswald P.E., Branscomb L.M. 2003, Valley of Death and Darwinian Seas: financing the invention to innovation transition in the United States, *Journal of Technology Transfer*, 28, 227–239.
[35] Knight F.H. 1921, *Risk, uncertainty and profit*, Hart Schaffner & Marx, Houghton Mifflin Co. New York.
[36] Scherer F.M. 1999, *New perspectives on economic growth and technological innovation*, Brookings Institution Press, Washington, DC.

3 Organizational Structures

3.1 ORGANIZATIONAL STRUCTURES FOR INNOVATION

The organizational structures for innovation consist of systems organizing knowledge, capital, and relations with the aim of developing new technologies. These structures are the result of the evolution of an activity that in past centuries was carried out essentially by individual inventors. The scientific developments and technological needs of industry already led, in the second half of the nineteenth century, to the formation of a first organizational structure for innovation that was later called research & development (R&D), based on a continuous generation of innovations, and not simply on development of a single inventive idea. The first R&D laboratories were created in Germany in 1870 by dye manufacturers [1] and research consisted in the synthesis of new molecules and verification of their suitability for dyes. During the same period, another industrial R&D laboratory was created in Germany by Carl Zeiss in 1876 for optical applications. Also in 1876, Thomas Edison created the famous research laboratory at Menlo Park, (New Jersey) with about twenty researchers. It should be noted that at the beginning of R&D activities, scientific knowledge was not necessarily fully exploited, with research often based only on large numbers of trials. This was the case of the Edison laboratory, where, for example, more than 1,000 materials were tested before finding the one that was suitable for electric lamps. At the beginning of the twentieth century, there was the creation of industrial R&D laboratories in the more developed countries. The United States had, for example, the Bell Telephone Laboratories of the American Telephone & Telegraph Company and the research laboratory of General Electric. Within the first twenty years of the twentieth century, 526 facilities for R&D were created in the United States [1]. Relations with universities were not particularly intense and limited mostly to consultancy. During the same period in the United States, new activity was born that consisted of supplying R&D services to industries with the creation of contract research laboratories. Such was the case of Arthur D. Little, founded in 1909 as a for-profit organization for contracting professional services including research, and the Mellon Institute, which started to sponsor industrial fellowship on a not-for-profit basis at the University of Pittsburgh in 1913 and became an independent industrial research group in 1927 [2]. In 1929, the Battelle Memorial Institute created a not-for-profit contract research laboratory at Columbus, Ohio with twenty researchers. After World War II, this institute had a great development with the creation of research centres in Europe at Frankfurt and Geneva. Interestingly, the Battelle Memorial Institute was established in 1927 by the will of Gordon Battelle, an Ohio industrialist who died in 1925. His idea was to create a bridge between research and industry, being unsatisfied with his experience with universities for the solution of a problem on treatment of zinc ore coming from one of his mines [2]. During World War II, the government three main R&D laboratories in the United States for the development of nuclear weapons: at Los Alamos, at Hanford, and at Oak Ridge [3]. Such a development, carried out between 1942 and 1946, corresponds to what is probably the greatest R&D project ever done; it was called the Manhattan Project, involving a total of about 130.000 people at the cost of about 26 billion US$ at the present value of the dollar. These laboratories continued their activity after the War, also exploiting nuclear research results for civil applications, and two further new important national laboratories were created at Argonne (Chicago) and Brookhaven (New York). The US government, around the year 2000, contracted the management of Oak Ridge and Brookhaven National Laboratories, as well as some minor national laboratories, to Battelle, which probably led to the largest research organization in the world with about 22,000 employees with a budget over six billion dollars. After World War II, there was a great expansion of industrial R&D laboratories in the United States, Europe, and even in Japan, while universities opened activities of contracted research with industry and creation of R&D laboratories. For example, the Stanford University created in 1945 the Stanford Research Institute, now SRI International, charging Battelle for its organization [2]. Such laboratories contributed later to the development of the Silicon Valley [4]. Until World War II contract research practically did not exist in Europe, but it developed rapidly. A study carried out by the European Commission in 1989 counted about 140 European contract research organizations with a total turnover of about one billion € at its present value [5]. Until the 1980s, industrial R&D was the main activity for technology innovation and was carried under conditions of competition and secrecy of development by various industries. After this time, many other nonindustrial actors became important for R&D activity, determining technical advances in industry [6]. In fact, technical innovations were the result of a variety activities, not only research in industrial R&D laboratories and contract research organizations, but also in independent private or public laboratories and universities, accompanied by cooperation among laboratories, trades of industrial properties, and search of competences in what has been called *distributed innovation* [7] as well as by new business model developments in what is called the system of *open innovation* [8]. It was in this frame that a new organizational structure for innovation was born, consisting of the SVC system, and characterized by a new radical strategy of financing and integration of R&D activities with business model developments. This structure has found a great diffusion especially in the Silicon Valley since the 1970s. Following the development that occurred in the last decades of the twentieth century in the field of ICT, and application of these new technologies in the social and economic fields, a new system of continuous relations among producers and consumers called *platform* [9] was born, evolving at the beginning of the twenty-first century as a new system of organizational structure, called *industrial platform*, based on a continuous relation in the offer and demand of technologies in the frame of an articulated system of development of innovations, and starting its diffusion especially for the introduction of ICT in the manufacturing industry. These three types of organizational structures, R&D, SVC and industrial platform systems, shall not be considered alternative but evolutive covering various types of innovations and markets. In fact, R&D activities are present in technological start-ups of the SVC system, and start-ups may be present in the industrial platform system.

3.2 THE R&D ORGANIZATIONAL SYSTEM

Technology dynamics considers R&D as an activity that organizes fluxes of capital and knowledge with the aim of developing new technologies. This definition is at the base of a model of the R&D process [10] and it takes inspiration from a previous model that was based on a domain view of R&D existing in industrial research laboratories [11] following the domain idea in organizations [12]. In this model for industrial research laboratories, R&D activity produces internal outputs in such contexts as reports, products, and processes, and external outputs such contexts as, publications and patents. Internal inputs for R&D come from, for example, marketing, manufacturing, and corporate strategies, while external inputs concern, for example, knowledge, and state of the art. Looking at this model, external inputs of state of the art, environment, combined with external output of presentations, publications, and patents may be considered as a flux of knowledge, while internal inputs coming from marketing, manufacturing, corporate strategies combined with internal output of reports, products, and processes may be considered as linked to a flux of investments or capital. On the other hand, R&D activity may not be considered as a simple black box, but as constituted by a system carrying out a set of R&D projects; this is at the origin of the technology dynamics model that considers R&D as an organizing activity of fluxes of knowledge and capital, in which the input for R&D activities is constituted by R&D projects that are determined by R&D investments financing R&D project proposals. R&D activity has output new technologies and has an available general knowledge that, combined with scientific, technical, and other external knowledge, generates new, innovative ideas and consequent new proposals for R&D projects. On the other hand, new technologies generated by R&D enter into use in the socio-economic system, possibly together with new, externally acquired technologies. The use of new technologies will require capital and is a source of possible return for investments (ROI). Yet, the socio-economic system using technologies determines new private investments, and possibly, also public financing for R&D, constituting the total R&D investment available in a territory. In Figure 3.1 we report a schematic view of the model of the R&D process in which the R&D activity starts with financed R&D projects selected among various proposals of R&D projects. Such proposals are prepared on the basis of innovative ideas generated by available knowledge coming from R&D activities combined with technical and scientific knowledge, possibly including newly discovered phenomena, other types of knowledge (for example, economic or environmental). R&D activity has an output constituted by new technologies and general knowledge, coming from either successful or abandoned projects and available for new project proposals. New technologies enter into use in the socio-economic system,

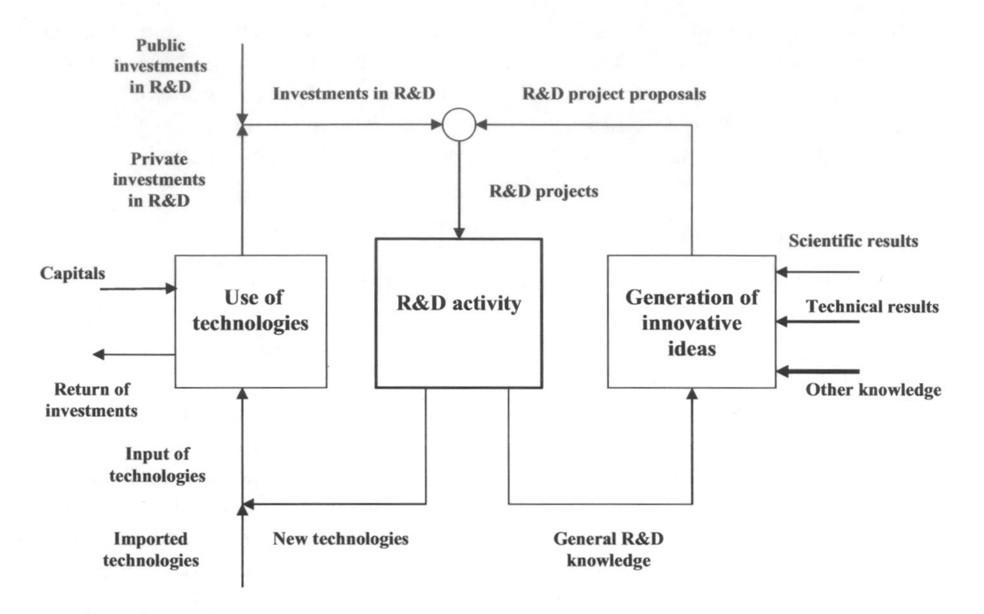

FIGURE 3.1 Model of R&D Process.

possibly also with imported technologies, requiring industrial capitals and generating ROI. The socio-economic system determines available new R&D investments, selecting R&D projects to be financed and to close the R&D cycle. The model is then composed of two fluxes, the first concerning capitals and the second based on knowledge; this last flux is studied in detail by technology dynamics, particularly assuming that R&D has a dynamic based on activities of projects. These projects generate a knowledge that has a key role in the development of further R&D project proposals.

3.2.1 Dynamics of R&D Activity

The model considers R&D activity to be composed of a set of various projects or activities that may be assimilated to projects, each defined, following the rules of project management, as a single, not repetitive, enterprise undertaken to achieve planned results within time and budget limits. The project's activities, and not its investments, are considered determinant for the R&D dynamics from the technological point of view. However, it is important to consider that there is a relation between project costs and R&D investments in a territory, as well as total cost of development of a new technology. Cost of project activity and R&D investments made in a territory in a certain period of time are represented by the following equation:

$$\mathbf{I}_{R\&D} = \sum_{i=1}^{N} \mathbf{p}_i \tag{1}$$

$\mathbf{I}_{R\&D}$ represents the total investments in R&D occurring in a certain period of time in a territory, while **p** represents the cost (investment) of each project of the N projects that have been carried out in the same period in the territory. From the point of view of the R&D activity, the costs of projects considered in Equation (1) may be present in various situations such as:

- Cost of projects terminated in the period but started in previous periods
- Cost of projects started and terminated in the same period
- Cost of projects started in the period but terminated in successive periods
- Cost of projects started in previous periods and terminated in successive periods

In Figure 3.2 we schematically reported the various possible dynamic situations of R&D projects in a territory and in a certain period of time. Another important aspect of R&D projects concerns the cost (or investment) for the development of a new technology. In fact, this development implies normally a sequence of projects covering the various stages of the innovation process until the moment in which the new technology enters into use or its development is

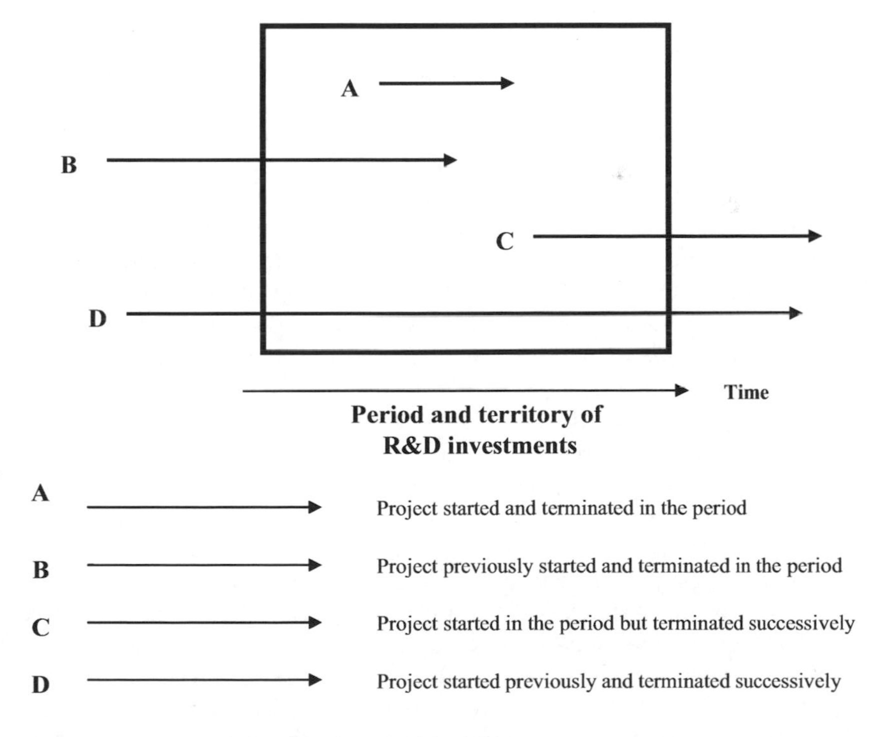

FIGURE 3.2 Projects Dynamics in R&D Activities.

abandoned. This means the cost (investment) of development I_{NT} may be expressed as:

$$\mathbf{I}_{NT} = \sum_{i=1}^{n} \mathbf{p}_i \qquad (2)$$

$\mathbf{I}_{NT}$ represents the total investment carried out in the development of a technology, while $\mathbf{p}_i$ represents the cost (investment) of each project of the n projects carried out for development. Such a sequence of n projects is, in most cases, extended through various periods of time (years) considered for R&D investments. It may be noted that the type of activity of the projects in the sequence follows normally the steps sequence of the innovation process reported in Figure 2.7. Actually, in the reality of R&D activity, only a minor portion of projects are carried out in a complete sequence during development of a new technology; many project sequences are discontinued, abandoning the development of a specific new technology. Consequently, considering a particular period of time and territory in R&D dynamics, the following situations occur:

- Projects that are abandoned in the period that stop .the development of the new technology
- Projects that continue beyond the considered period of time
- Projects that are terminated in the period and generate new technologies entering into use

In the first case, discontinued projects generate, from the economic point of view, a financial loss involving not only costs of these projects, but also those of previous projects of the sequence. In the second case, returns on investment and growth appear only after the sequence of projects completed in a further period of time. Only in the third case could the completion of the sequence of projects trigger the generation of economic returns through the entering into use of a new technology. It should be noted, however, that projects that are terminated successfully in the considered period may generate returns on investments and growth only in successive periods, and are related to previous R&D investments. In conclusion, *any direct relation between the R&D investments and economic growth of a territory recorded in the same period of time does not actually exist* but the existence of a time lag between R&D investment and the moment a new technology may enter into use, and a second time lag between this moment and the periods in which the new technology generates return of investments and growth. Such time lags should be considered when the relations between R&D investments and growth are studied. Another important aspect of R&D dynamics is the role in the model of each project that constitutes the sequence leading to a new technology or to the termination of development. In fact, at the end of each of these projects in sequence, there is an evaluation of results that leads to the termination of development, in the case of unsatisfactory results, or in the case in which the technology is ready to enter into use. In the case of satisfactory results, but not termination of development, a new proposal for a project continuing the development is prepared, with preparation occurring in the frame of the flux of knowledge of the R&D model. Such proposals may be accepted or refused following the strategies and selection criteria of the entities financing the development of the technology.

3.2.2 Fluxes of Knowledge and Capitals in the R&D Model

The R&D model describes a flux of knowledge that includes: the generation of innovative ideas by combining R&D knowledge and external knowledge (scientific, technical, etc.), forming proposals for R&D projects and confronting proposals with available investments for R&D. Furthermore, the financed R&D projects are carried out in the frame of a certain number of steps characterizing the innovation process. As previously noted, the technological model of R&D cannot explain either the various aspects of flux of capitals or the socio-economic factors that may favor or inhibit the process of generation of new ideas and R&D proposals. Neither could it define criteria for selection of proposals for R&D projects; establish objectives, performances, or economics of the developing technology, finally decide the continuation or discontinuation of development or whether to enter the new technology into use. We may, however, give a technological description of the various steps occurring in the flux of knowledge and flux of financing R&D projects.

3.2.2.1 Generation of innovative ideas

The model considers generation of innovative ideas as the result of a combinatory process involving knowledge generated by R&D and knowledge of a scientific and technical nature, as well as other knowledge that may be of, for example, economic, market, or environmental in nature. The process of generation of innovative ideas by exploiting scientific results, or simply by combination of previous technologies, as well as generative aspects concerning individuals or group of individuals has been described in Chapter 2, dedicated to the process of technology innovation. In this description are emphasized creativity and the entrepreneurial attitude that favors the birth of innovative ideas, which are also valid for R&D activities in which new ideas are obtained from knowledge generated by R&D activity combined with external knowledge of technical, scientific, or other nature.

3.2.2.2 Elaboration of R&D project proposals

In order to start the development of an innovative ideas through an R&D process, it is necessary to elaborate project proposals that establish, objectives, research programs, timing and budgets necessary for projects submitted for financing. If we consider a territory in which a certain amount of knowledge is available that is useful for the preparation of R&D project proposals, we may define the efficiency of a territorial innovative system (ISE) as the capability of the territory to exploit the available knowledge that generates innovative ideas for R&D project proposals. This capability depends in a certain way on the efficiency of actors involved in R&D in exploiting knowledge through its free flow that exists within the system or externally. Such exploitation may be strongly motivated by various factors such as availability of financing, or inhibited by an unfavorable climate for innovation developments and lack of capital. Another aspect for R&D project proposals is the importance of collateral studies concerning economic, market and even pre-feasibility studies on the innovative idea that may improve the research program of the proposal and, in the case of contract research, all that makes the proposed R&D project more interesting for industrial financing.

3.2.2.3 Financing R&D projects

The model cannot enter into discussions either about capital availability for R&D financing, or selection criteria for the choice of proposals that should be supported. It limits the discussion on the level of available financing compared with the required budgets of the R&D project proposals, specifying the possible situations as follows:

- Limited availability of investments for R&D with respect to overall budget proposals for R&D projects. This is the most common case. Availability of R&D investments and selection criteria are not normally related to proposed R&D project budgets but depend on various socio-economic factors.
- Scarcity of proposals with respect to available R&D investments (typically public financing) observed, for example, in poorly developed or industrially declining territories.
- Full availability of R&D investments to cover all valid proposals that may lead to an exponential growth of financed R&D. This case, which is much less frequent, may be present in periods of war and in certain regions and phases of development of new technological sectors. A great example of full availability of R&D investments is represented by the Silicon Valley in which donation, industrial capitals and VC make available investments for almost all valid research proposals.

It should be noted that, following the model, the exponential growth of innovative ideas and then R&D projects, when enough financing is available, is essentially the consequence of the combinatory nature of generation of new ideas, based on increase of R&D knowledge due to previous R&D activity in a typical autocatalytic process of increasing returns on innovative ideas.

3.2.3 Role of the General Knowledge Generated by R&D Activity

Following the model, the R&D activity generates, besides possible new technologies, an R&D knowledge (RDK) of a general nature that has an important role as driving force for the development of new R&D activities. Such knowledge is generated by either successful or abandoned projects that produce new ideas not necessarily related to the technological objectives of these projects. RDK does not only include technical, scientific, or other information, but also specific know-how formed in the various research fields because of the technological nature of R&D activity. As technological innovations are the results of combinatory processes [13], possibly exploiting scientific results [14], the RDK may be considered a knowledge available for use in the combinatory process with pre-existing technologies, generating a new technological idea. RDK is not in itself patentable differently from derived technologies. It should be noted that RDK is subject, with time, to fading effects and then a loss of knowledge. This loss may be higher in the case of industry when part of generated RDK might be not of interest in evolving technological strategies of firms. RDK does not correspond to knowledge spillover, used in many economic studies on R&D, defined as an externality of R&D in a firm that impacts the technological innovation in other firms [15], and found effective in boosting growth in industrial and developing countries [16]. Such defined knowledge spillover does not take account of knowledge generated outside firms' activities that do not necessarily have economic purposes and by other R&D actors that exist besides firms (universities, public and private laboratories, contract research organizations and start-ups), in a distributed innovation system [7] and in the frame of an open innovation environment [8].

The process with which the RDK is diffused, exchanged, and exploited for generation of new technologies in a territory is quite complex and varied. We illustrate this by presenting a number of real cases. The first case concerns the way with which RDK is generated and exploited successfully in the Silicon Valley. In the second case, we treat the generation of RDK in a large R&D project for military purposes, such as in the Manhattan Project. The third case covers the diffusion of RDK among such firms as Xerox, Apple, and Microsoft for graphical interfaces for screen bitmapping might be considered a case of knowledge spillover. The fourth case treats a technology developed for pharmaceutical purposes through a spinoff and creation of a start-up in a contract research laboratory for metallurgical applications. Finally, the last case concerns a technological operation developed for nuclear energy production that migrated in various other technologies in such sectors as surface hardening, steel refining, and production of nanoparticles, the last two cases being a direct experience of the author. The Silicon Valley innovative system and role of RDK may be described as follows:

> In the Silicon Valley an initial idea for a new technology is generated in manifold ways and considered more as a trigger of innovations than a potentially patentable invention. These ideas are freely discussed and improved constituting in what it may be considered an informal exchange of RDK. On the other side RDK formed internally to firms is continually exchanged in the frame of their activities in form of projects for new products through cooperation, dismissing or hiring following the need of the projects supported by the existence of a local market of competences. In this case a person may be dismissed or hired, as a function of his competence, several times independently he has worked in the meantime for a concurrent firm favouring in this way RDK exchanges. It should be noted that such type of RDK does not necessarily arise patents conflicts as it represents often a knowledge originated by projects, but not necessarily linked in generating new ideas to the objectives

of the projects in which it is formed. The fact that useful innovations may be generated by projects carried out for other purposes, justifies the existence for example of specific budgets in contract research organizations available for development of such collateral ideas. These budgets exist also in universities with strong links with industry such as Berkeley and Stanford. It should be noted that Silicon Valley found a great support by military research during the war of Korea and cold war developing in particular integrated circuits and mini-computers technologies winning the competition with electronic industry of Route 128 near Boston also by a better exploitation of RDK exchange among firms against closure existing in electronic industries of the Boston area [17]. Exploitation of RDK for civil applications derived by this military R&D activity were considered only after some détente of cold war and oil price shock at beginning of seventies [4] boosted with the arrival of PC followed by an enormous development of ICT.

The Manhattan Project for development of nuclear weapons has been described in detail in a book of Richard Rhodes on the history of making the atomic bomb [3]:

In order to develop nuclear weapons, the American government created in 1942 three main research laboratories at Los Alamos, Oak Ridge and Hanford. However only the Los Alamos laboratories charged to build up the bomb was operated directly by the Army, the other laboratories, Oak Ridge for the production of uranium and Hanford for the production of plutonium, were given for operative management to two main US chemical industries respectively Union Carbide and Du Pont with the purpose to exploit their technological knowhow useful for the project but also to exploit research results by these industries for future civil applications with economic returns. At the end of the project Du Pont ceded management of Hanford to General Electric, however this company was interested, for production of energy, to develop pressurized or boiling water nuclear reactors instead of graphite reactors used in Hanford for plutonium production, and the Pacific Northwest National laboratory, built near the Hanford site for civil nuclear applications, was finally ceded in 1964 to Battelle, while Hanford site remained under the control of the Atomic Energy Commission. Union Carbide operated Oak Ridge longer until its disappearance after the Bhopal disaster occurred in 1984 and Oak Ridge operation was finally ceded by government with other minor national laboratories in 2000 to Battelle. Nuclear research carried out during the war generated a great number of civil technologies from nuclear reactors for production of energy to fluorinated plastics such as Teflon® used for example for frying pans with an anti-adherent surface for cooking.

The case of RDK diffusion from Xerox to Apple and Microsoft with respect to user interfacing by screen bitmapping, a milestone in PC development and presently also used in tablets and smartphones, is sometimes described as the biggest heist in the chronicles of industry and has been described in detail in the official biography of Steve Jobs [18].

The idea of development of a graphic user interface using screen bitmapping was born at Palo Alto Research Centre (PARC) of Xerox company in the second half of seventies. In1979 Xerox's VC was interested to invest in Apple and Steve Jobs agreed to the buying of 100,000 shares at a convenient price of 10 US$ if Xerox

> opened information about R&D activity at PARC. Xerox accepted and during a series of meetings at PARC it was presented a total of three interesting features of its research concerning: networking of computers, object-oriented programming and graphic interface for screen bitmapping. This last feature, used at PARC for the moment only in computer prototypes, was found of great interest by Steve Jobs that started R&D at Apple on this subject. Apple's engineers in fact significantly improved the graphical interface that was applied for the first time to the successful PC model Macintosh, launched in 1983. Xerox tried also to launch in 1981, well before the Macintosh, the Xerox Star computer using a graphic interface, but it was expensive and less performing than the Macintosh and it was a commercial failure and, after that, Xerox abandoned this line of products. Later the graphic interface developed by Apple interested Microsoft for its software development taking advantage of acquired knowledge in the collaboration with Apple on this topic. In fact, Microsoft had an agreement with Apple that it will not create graphic interface for anyone other than Apple until a year after Macintosh commercialization. Unfortunately for Apple the agreement did not provide that commercialization of Macintosh would be delayed for a year. In this situation Bill Gates thought that was his right to reveal that Microsoft planned to develop a new operating system, based on knowledge obtained during collaboration with Apple, called Windows, for IBM PCs featuring a graphical interface. Of course Steve Jobs was furious because of this announcement but legal actions were unsuccessful, and Bill Gates commented litigation with Steve Jobs in this way "It is more like we both had this rich neighbour named Xerox and I broke into his house to steal the TV set and found out that you had already stolen it."

The fourth case concerns a spin-off exploiting RDK generated in a great contract research laboratory, specifically a technology applied successfully in fields completely different from that of the original project, and shows how contract research organizations may induce exploitation of RDK more efficiently than industry:

> In the middle of seventies an important Swiss pharmaceutical company contracted with the Battelle Research Centre in Geneva a R&D project on development of a granulation technology based on exploitation of effect of ultrasound vibrations on a laminar jet of liquid. Through a suitable tuning there is the appearance of vibrated nodes in the jet that, after solidification, form spherical and identical granules. In this way it is possible to dissolve a pharmaceutical principle in a fusible bulking agent obtaining a drug in form of granules of identical dimension and composition. Although good technical results, after few years, this project was discontinued in favor of other granulation technologies. However, a discussion internally to the research center between a micro-granulation expert and a metallurgist made the birth of an innovative idea to use the same technology to granulate special metals that cannot be grinded because too soft, in particular calcium metal that had at that time growing applications in form of granules with the same dimension. These two researchers made a spinoff in 1982 creating a company in France named Extramet SA at Annemasse, a city near Geneva, with own capital and public aid proving the feasibility of granulation of calcium metal and finding French VC to terminate development and to build up an industrial production plant entering in function few years later with the firm name of Extramet Industries, changed in 1992 in IPS. After calcium granules production the technology was extended to granulation of tin soldering alloys used in printed

circuits for electronic applications. This company reached a dimension of about 30 employees but finally it was closed in 2017 after more than 30 years of activity. If the pharmaceutical company would have made granulation research in its own laboratories, after discontinuation of the project, researchers, not aware of metallurgical art, would be assigned probably to other tasks and RDK of this research lost. Now supposing a metallurgist that has the idea to use ultrasounds in the same way to form metallic granules, he would probably succeed but without availability of previous RDK this development would take more time and have higher costs.

The last case presents a technological operation developed for a specific technology in nuclear energy production that has been transferred to a series of completely different technologies, showing how RDK may diffuse in the interconnected network of the technology ecosystem:

At the beginning of seventies of the past century Battelle Research Centre in Geneva was developing a new product for hardening steel based on calcium carbide dissolved in molten salts as alternative to toxic and polluting molten cyanides used currently. The team faced a problem how to industrially discharge the molten product from the dissolution reactor. Pouring operation, largely used for molten metals, was not suitable because of destruction of contained calcium carbide in contact with air at high temperature. On the other side a mechanical valve operating around 800°C was not considered technically feasible. The team was in contact with many researchers involved in scientific research and technical use of molten salts and had the occasion to obtain a report on component systems of the molten salt fast breeder nuclear reactor, developed and later abandoned, at the Oak Ridge National Laboratory [19]. The nuclear reactor operated with a fuel based on uranium hexafluoride dissolved in molten salts and had the necessity to evacuate rapidly the melt from the reactor. For this purpose, it was developed a freeze valve constituted simply by an open tube at the bottom of the reactor in which the molten salt cooling forms a solid plug for containment of the melt. For evacuation an auxiliary furnace at the level of the plug melted the salt resulting in a rapid flow down of the melt. This freeze valve was applied successfully to discharge of molten product for steel hardening, and even applied in other similar productions of dissolved calcium carbide in molten salts for pig iron refining and dissolved calcium metal for steel refining. The same freeze valve was also used for production of advanced ceramics, composed by nanoparticles of titanium carbide, synthesized in a molten salt reactor developed by the previously cited Extramet. All these technologies were developed at the stage of small industrial production plants but later abandoned because not competitive.

All these real examples show well the great variety of processes that assure diffusion of RDK for the generation of technology innovations, as well as the importance of R&D that is carried out not necessarily for economic purposes, and the possible higher efficiency in exploitation of RDK for R&D projects in research organizations instead of in industrial R&D laboratories.

3.2.4 Interface between R&D Activity and Scientific Research

Scientific research is an important source of phenomena exploitable for applications in new technologies [14] and it is also recognized as a source of knowledge useful in technical searches to improving existing technologies [20]. It is commonly considered that scientific research is at the origin of, and then precedes R&D activities; however, in certain cases, R&D needs specific scientific knowledge for its purpose, but is not already available, inducing the carrying out of scientific research to obtain such knowledge. The fact that R&D may also induce scientific research makes the relation between science and technology quite complex and in developing R&D activities, especially in new fields, there is often an intertwining process between R&D and supporting scientific research during projects activity making it difficult to separate what is done for scientific from what is done for applied purposes. Such intertwining between scientific research and R&D has existed since the beginning of R&D activities in the nineteenth century. Searches for new dyes in the first R&D laboratories involved synthesis of new molecules, a typical scientific activity, and verification of their interest as dyes, a typical R&D activity. Technology dynamics suggests clearly distinguishing scientific research as an activity improving knowledge of nature that might be the potential object of scientific publications, from R&D project activities that verify the use of phenomena for technological purposes that might be the potential object of patents. Actually, there are many examples of scientists who gave important contributions to patented inventions through both research and R&D activities. It is even known, for example, that a patent was obtained for a pump for water in 1594 by Galileo Galilei, father of modern science, under the law of the Republic of Venice. Another important case concerns the activity of two scientists, Leo Szilard and Enrico Fermi, regarding exploitation of nuclear fission of uranium for energy production, a case that is described in detail in a book about the making of the atomic bomb [3] and that merits to be reported:

> Enrico Fermi, a famous Italian physicist, studied the effect of bombarding various atoms with neutrons in the Panisperna laboratory in Rome. In fact, he tested also uranium but he did not recognize the fission effect that actually was discovered in Germany by Otto Hahn and Fritz Strassmann in 1938. It is interesting to know that the possible use of this effect in production of energy was anticipated by Leo Szilard, a Hungarian physicist friend of Albert Einstein. Incidentally it may be noted that Szilard worked together with Einstein at the Kaiser Wilhelm Institute in Berlin on refrigerator technologies for the German firm AEG obtaining together a certain number of patents. Emigrated in UK, Szilard, aware of effects of neutrons on atoms in generation of energy, he thought that if in the collision there was emission of an increased number of neutrons there was the formation of a chain reaction with the production of an enormous amount of energy that might have also a military use. In 1936, two years before the discovery of uranium fission, he was able to obtain a patent on this idea convincing the British Admiralty to keep it in secrecy. Enrico Fermi, after the obtention of the Nobel Prize in 1938, emigrated in USA teaching at the Columbia and at the Chicago University and he had the idea to build up a reactor generating energy by controlling the fission of uranium. In Chicago he designed and built up a system constituted by a pile of graphite and uranium bricks. The graphite used as moderator to slow down neutrons in order to increase the collisions with uranium. He made available also cadmium rods to be introduced into the pile in order to stop the fission by using the property of cadmium in the absorption of neutrons. The pile

> was monitored with instruments measuring the emission of neutrons and, on December 2, 1942, the pile reached the critical condition of generation of energy. Fermi interrupted promptly the reaction just after 4.5 minutes at 1.5 watt of power otherwise, as a cooling system was not present, an uncontrolled functioning for more than one hour, would result in the development of millions of kilowatts generating a little Chernobyl disaster in Chicago. In fact, the Fermi reactor was not used for production of energy but for the production of plutonium, another fissile element, by-product of effect of neutrons on uranium. A certain number of reactors were built up at the Hanford National Laboratories for this purpose. Plutonium was used in the first nuclear bomb experiment in the desert of New Mexico and for bombing Nagasaki. After the war the Atomic Energy Commission paid fifty thousand US$ to Enrico Fermi and to Leo Szilard for their inventions related to generation of nuclear energy and to the atomic bomb [3].

This historical example shows well the process by which scientific results may be exploited for new technologies depends greatly on an open view for applications by researchers. This fact is at the origin of a gap of efficiency in exploiting scientific results between Europe and the United States that is characterized by a more cultural view of scientific research in Europe, while in the United States, the entrepreneurial view is historically prevalent [21]. Such entrepreneurial gaps in Europe have also been recently confirmed by studies carried out, for example, in Italy [22] and in a certain measure, also in the United Kingdom [23].

3.2.5 Interface between R&D Activity and the Socio-economic System

The main aspect that exists at the interface of R&D activity and the socio-economic system concerns the relation between investments in R&D and the economic growth of a country. This argument will be treated in detail subsequently in a discussion of a mathematical simulation model of R&D and in the chapter on the applications of technology dynamics. We treat two minor arguments concerning competition between endogenous technologies developed by R&D activities versus imported technologies, and differences in efficiency of private investments versus public investments or aids for R&D activities, both involving technological aspects.

3.2.5.1 Endogenous vs. Imported Technologies

The role of imported versus endogenous new technologies is particularly important in emerging countries supporting their industrialization by policies concerning either R&D activities or import of technologies. The effects of such policies have been studied, for example, confronting economies essentially based on import of technologies in the case of GCC countries, such as Saudi Arabia, BRIC countries based on R&D, such as China, showing that an innovation-based economy actually is not dependent on total expenditure on R&D, but rather relies on the efficient allocation of investments and the rigorous implementation of innovation strategy [24]. For our R&D model, there are important differences in these two strategies concerning generated RDK, which is nearly absent in the case of imported technologies; by consequence, this policy sensibly limits the future possibility of endogenous generation of new technologies, especially those that have a high radical degree and are thus competitive.

3.2.5.2 Private vs. Public R&D Investments

The differences in efficiency between private and public R&D investments have been studied by many authors [25] who found, in most cases, higher returns in R&D in countries where the ratio between private and public financing is high. It has sometimes been observed that, when public investments are prevalent, there is a negative consequence for gross domestic product (GDP) [25]. Many explanations have been advanced for the major efficiency of private versus public R&D investments, including the fact that public investments generally involve R&D projects with higher uncertainty and greater risks of failure, and that industry has more experience and knowledge in carrying out R&D projects [26]. The R&D model cannot enter into such discussions, but offers another point of view indicating two critical points in the innovation process in which uncertainty and need of R&D investments are both high and merit consideration for public aids. The first point concerns the phase of generation of innovative ideas and R&D proposals that would be better followed by preliminary economic and market studies and even prefeasibility studies previously mentioned. The second point is that, in the final part of development, high R&D investments are necessary, for example, in construction of pilot or demonstration plants or production of prototypes, but nevertheless in conditions of high uncertainty of success. It should be argued whether direct public aid to industry for R&D is an effective way to support a growth. For example, Switzerland does not supply any direct aid to industry for R&D; nevertheless it is considered one of the industrialized countries with the best results in technology transfer [27]. In fact, in Switzerland, differently from many European countries, public aids for R&D are given to polytechnics and technical universities in support of search and implementation of industrial research contracts, and firms receive aid that is limited to industrialization projects. As shown by the model, R&D in polytechnics and technical universities would be made particularly effective by improving exploitation of RDK generated during contract research with industry for future project development.

3.2.6 Mathematical Simulation of the R&D Process

It is generally recognized that R&D activity is an important factor for determining the socio-economic growth of a territory. It is then of tentative interest to find a quantitative relation between R&D investments in a territory and its growth through the presented R&D model. Using the model, it would be possible to study the effect of starting a certain number of R&D projects in a territory and the possibility of generating a certain number of new successful technologies to trigger socio-economic growth [28]. In this simulation of R&D activity, *the number of financed R&D projects, and the number of generated successful technologies, simulate the magnitude of R&D investments and that of economic growth, respectively.* This simulation needs the estimation of various parameters that concern the number of R&D project proposals generated by available RDK and additional external knowledge, depending on the territorial efficiency in exploiting knowledge in the generation of innovative ideas, as well as a certain number of rates to determine the selection of financed R&D projects, the generation of new technologies, and the number of successful technologies defined as new technologies reaching an elevated level of activity and return on investments. The mathematical description of the model with the definition of the various parameters and the estimation of variables used in running the model are reported in detail in Appendix 2 at the end of this book. We present the detailed qualitative aspects of the model and its results. The aim is to get an idea of conditions in which a certain amount of R&D investments, represented by an initial number of R&D projects carried out in a territory, might succeed in the generation of successful new technologies and then economic growth. R&D is a quite complex process and, in our attempt, we have considered, in fact, only a rough simulation of such a process in what may be considered a toy model, but that, however, is a measure to get results of interest. Furthermore, many parameters necessary to quantitatively operate the model, are, in fact, not really available from statistical data and should be substituted by reasonable, but only indicative, values suggested by experience in R&D activity. The running of the model consists of determining the effect of a variable number of initial R&D projects on possible generation of new and successful technologies, and considering exploitation of RDK and external knowledge, as a function of efficiency of the territory in the generation of a further number of R&D projects proposals. We have assumed in these calculations that there are always enough R&D investments to cover budgets of all valid proposals generated by the innovation system of the territory. We suppose also that the innovative system of the territory operates by a more or less advanced regime of open innovation, and that technology innovation is carried out following a more or less advanced distributed innovation system. This means that actors carrying out R&D projects are not limited to industrial R&D laboratories but also to, for example, contract research organizations, private or public research laboratories, start-ups, exchanging in a certain measure the generated RDK in the territory [10]. The R&D activity is seen in term of projects and, although normally a new technology is generated by a sequence of R&D projects, for simplification in operation of the model, we consider that a new technology may be generated by a single project. Actually, generation of R&D project proposals, starting of new R&D projects, and formation of new technologies occur continually; however, in running our model, for simplification of computation, we consider R&D activity in term of cycles, each fed by a certain number of R&D projects, and generating or not generating new technologies. This means that the duration of generation of R&D project proposals followed by carrying out R&D projects is the same for all projects and equal to the cycle time. The mathematical model should also indicate how to calculate the number of generated R&D project proposals from available knowledge, taking account of the combinatory nature of the generation of innovative ideas. For this purpose, the mathematical model considers knowledge in terms of information packages, and each successful or abandoned project generates a certain number of information packages. The total of generated packages from R&D is further increased by a certain amount of information packages coming externally from scientific, technical, or other knowledge. However, as RDK may be partly lost with time by a fading effect, we have assumed a percentage of loss of knowledge, then of the number of information packages occurring at each cycle and concerning all remaining knowledge of previous cycles. The potential generation of innovative ideas is, following the model, the result of combination of a few available packages. It is then possible to define such number and, by combinatory calculation, to obtain the total number of potential combinations from the total of calculated available packages. The number of generated R&D project proposals would result from a selection among all available potential combinations determined by the ISE of the territory, defined as the capabilities of the R&D activity of the territory to generate R&D project proposals in respect to the available knowledge in term of RDK and external knowledge. Higher is the value of ISE, higher would be the number of available R&D project proposals. Concerning the socio-economic growth of the territory we have considered, in a simplified view, that it is proportional to the resulting number of new successful technologies that are in fact a part of all new technologies entering into use. It is possible with the mathematical model to study the effects of an increasing number of initial R&D projects, equivalent to initial R&D investments, on the number of generated new technologies, and possibly successful technologies, equivalent to economic growth, as a function of ISE. The results obtained are represented schematically in Figure 3.3 in which there are two curves, elaborated with respect to the results after a determined number of cycles, the lower curve representing the total initial number of R&D projects necessary to generate at least one new technology, the upper representing the same for the generation of at least one successful technology, both as a function of the values of the ISE of the territory. The results, in

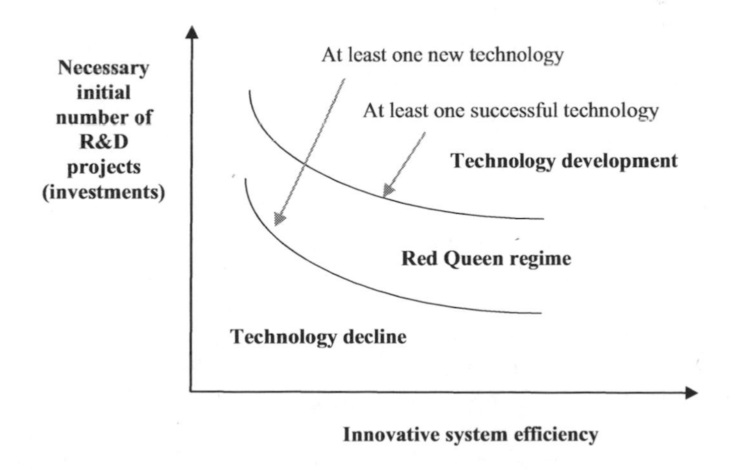

FIGURE 3.3 Obtained Regimes of Technology Innovation as a Function of ISE.

fact, separate the graphic into three areas: the first one below the lower curve corresponding to the absence of generation of new technologies and indicating a situation of technological decline of the territory; the second one above the upper curve corresponding to the generation of successful technologies and then to a situation of technological development of the territory; and finally the area between the two curves corresponding to generation of new but not successful technologies, indicating a situation of stagnation of technology development typically represented by a Red Queen regime (described in Chapter 2 "Technology"). It may be noted that the increase of the ISE of the territory has the effect of a decrease in the minimum amount of the initial number of R&D projects necessary to generate a new or a successful technology. This means that for the same amount of R&D investments, it is possible to obtain a greater economic growth of a territory characterized by a high value of ISE and showing the importance of exploitation of available knowledge. The conclusion obtained by this mathematical model is that *economic growth of a territory does not actually depend on R&D investments, that should rather be considered a means, but on the intensity of generation of innovative ideas, which depends on the efficiency of exploitation of knowledge by the territory, and of course, by adopted strategies and availability of capitals financing technology developments.*

3.3 THE START-UP – VENTURE CAPITAL SYSTEM

The start-up – venture capital (SVC) system represents another way to make technological innovations resulting from evolution of industrial R&D activities toward a distributed technological innovation system [7], accompanied by development of new business models in which it is called a regime of open innovation [8]. In fact, the first venture capital (VC) company already appeared just after World War II in the United States [29], but a great development of this type of financing innovations has only occurred since the 1970s, especially in territories such as the Silicon Valley, with start-ups becoming an alternative to industrially financed R&D projects in developing technological innovations. Actually, it appears that this system's success is based more on a new radical approach introduced by the VC in financing innovations, than on activity of start-ups. Industrial capitals financing R&D projects normally have the objective of obtaining an ROI that exploits the use of new technologies. The innovation of VC has overturned this objective by financing innovation, not by R&D projects but by start-ups, being disinterested in exploiting the new technologies, but selling the developed technologies and refinancing with resulting ROI new technology developments. In this way, after retention of part of ROI as reward by VC and possibly, in part, also made available to the start-up founders, when ROI is largely positive, an autocatalytic cycle of investments and reinvestments is formed with the remaining ROI that, in technological fields with large innovative potential attracting capitals, as occurs in the Silicon Valley, leads to a great development. In fact, the SVC system has already been the object of some studies concerning its general technological aspects: interfacing, financing, and technological risk management, especially considering start-ups having technological innovations as objectives and studying the comparison with the industrially financed R&D projects system [30].

3.3.1 Start-up Description

A start-up is a company of small dimension that begins its activity having, however, different characteristics from current small companies that are also beginning their activity. We principally describe here start-ups with technological objectives, and not those involved in the development of new technologies with socio-economic objectives, which are already cited in Chapter 2 ("Technology") but that do not follow the same development process. Start-ups with technological objectives described here are then involved in development of new technologies, and not in application of ICT in innovations of social or economic nature. Start-ups are typically financed by VC and not by industrial or public capitals. Although start-ups are organized legally as companies, in fact, their activity is more like that of a project defined as a nonrepetitive activity undertaken to achieve an objective within a time limit and budget that, in the case of start-ups, is derived by their capitalization. The objective of a start-up is reaching an exit that consists normally of selling of the developed technology, typically to a great company, or of collecting capital, for example, by being quoted in a stock exchange, in order to become an industrial company. With the exit of the start-up the VC obtains, by selling the technology, an ROI that is used for reinvestment in developing new start-ups. The development of the activity of a start-up follows various phases that are explained in the following paragraphs.

3.3.1.1 Start-ups generation and development

The activity of a start-up begins with the generation of its innovative idea followed by various stages of development until its exit. This process, in the conditions of the best practice for the development of technological innovations, is similar to the development of an R&D project, with the steps of generation of innovative ideas, feasibility and development until favorable conditions for an exit, or, on the contrary for abandonment of the start-up. However, start-up activity is not limited to R&D, but is accompanied by other activities, particularly the development of a business model suitable for the type of developed technology and possibly commercial activities. The fact that start-ups may have important R&D activities, many typical processes of R&D are present, particularly the generation of general knowledge, independently of its success, that constitutes, as in R&D activity, a driving force in generating new innovative ideas useful for new start-up projects. Like the case of industrially financed R&D projects, successful growth of start-up activity in a territory may be obtained only above some critical number of start-ups and VC investments. The generation of innovative ideas for the creation of start-ups is favored by an innovative and entrepreneurial climate with free discussion and exchange of ideas and creativity as typically exists in the Silicon Valley. The common initial process with which the innovative idea begins its realization is called *spin-off*. It consists of researchers or technicians leaving university structures, public or private research laboratories, and even industrial R&D laboratories after deciding to develop an idea derived from their work in an autonomous manner. Also typical is the spin-off from innovative firms, such as Fairchild Semiconductors, which, from 1959 to 1971, directly or indirectly generated 35 start-ups, reaching the present number of 92, which may be traced to this electronic industry of the Silicon Valley [31]. However, spin-off is not the only initial process in generation of a start-up, and sometimes their origin is due to people who pursuit an innovative idea, even they had already done so during their education, and decide to leave their employers who are not necessarily concerned by their innovative ideas.

3.3.1.2 Development of the model of business

The development of new business models represents an important aspect of technological and economical innovation in what it is called the regime of open innovation [8]. It is also an important activity for start-up development in order to match the developed technology with the best model of business for its exploitation to increase the possibilities of a favorable exit, either by selling the technology to industry or finding capitals for transformation in an industrial company based on the business model developed in accord with the technology. The term business model is widely used and the main functions of this activity are [32]:

- Identification of a market segment in which technology is useful and for what purpose
- Articulate the value created for users and structure of the value chain, i.e., network of firm's activity required to create and distribute products or services to customers
- Estimate cost structure and profit potential and formulate the competitive strategy by which the innovating technology would gain and hold advantage over rivals

In business models, the value thus derives from the structure of the situation, rather than from some inherent characteristic of the technology itself, and technical uncertainty is a function of market focus and varies with the dynamics of change in the marketplace [32].

3.3.1.3 Commercial development

During the final phases of development of a start-up, it is common to start commercial activities in terms of offers of products or services to the market. Such offers, however, in many cases function more in showing the validity of the technology to search further important financing, than in supplying financing to the start-up in the form of commercial revenues. In fact, market knowledge is considered of main importance and market risks considered superior to technical risks in the start-up activity [33]. Risk and uncertainty of market acceptance, as in R&D project developments, may be reduced sensibly only in the final phases by directly testing the market [34].

3.3.1.4 Structures and organizations promoting start-ups

Promoting structures for start-ups are of various types, following the various stages of development. In the beginning, coworking structures offer office emplacements in open spaces, meeting rooms, common equipment, such as printers and photocopiers, as well as meeting spaces with drinks, vending machines, or bar. A variant of coworking space is the open lab. It is a space with various equipment such as welding machines, tools, and 3D printers that are useful to make prototypes. For start-ups in an advanced phase, there are incubators or accelerators that offer office spaces or even spaces for small productions or laboratory work. For more advanced start-ups, there are scientific or technological parks with available small industrial buildings. Promoting organizations for start-up activities may be done by, for example, public or private entities or associations that have as their primary or secondary objective the promotion of start-ups with services such as coaching, mentoring, training and education, study tours, and technology scouting. Sometimes, their help includes searching of financing or even acting as a source of seed capital for the initial phases of start-ups.

The possibility also exists of creating start-ups based on a research laboratory activity with the aim of operating for either contract research with industry or for the generation of start-ups that exploit RDK generated by laboratory research. An example of this type of company is the Generics Group, located near Cambridge England, and founded in 1986 by Gordon Edge as a business consulting firm with a scientific multidisciplinary approach that uses profits to create value through investing in intellectual property. This company had flat structure typical of start-ups and, when enough cash was available, generated spinoffs with VC but with Generics maintaining an equity stake. This company developed a model of business, formalized in 1998, based on both technology and business consulting, as well as on licensing, investing, and creating new businesses. This activity, including fully equipped laboratories and availability of early internal funding for projects, reached a staff of nearly 250 employees in 2002. An earlier example of smaller dimension, but based essentially only on technology innovations developments, was the interesting case of Extramet SA, already cited as a case of diffusion of RDK from R&D activities, had this history:

> Extramet was founded by Gérard Bienvenu and Bernard Chaleat in 1982 with a spinoff from the Battelle Geneva Research Centre, and it was based at Annemasse in France in the proximity of Geneva. It reached a dimension of more than ten researchers and in ten years of activity it generated four startups with French VC concerning micro-granulation of metals, elimination of heavy metals pollution, production of advanced ceramics and metal powders refining in inductive plasma, but it had also important research contracts with industry for the production of calcium metal and titanium and even an important multiclient study concerning application of metallothermic processes with worldwide clients in Europe, USA and Australia. It ceased its activity in 1992 because of lack of capitals financing startups in which it was involved, however, three of its research engineers, leaving the company, founded other three new startups. Furthermore, one of the founders, Gérard Bienvenu, continued an activity of creation of startups founding Easyl in the field of advanced ceramics, and Gerkaro Sciences in the field of

energy storage leading successfully to generation of two subsidiaries Hevatech and Ergosup.

The history of Extramet shows well the importance of previous experience in contract research and start-up activities in promoting entrepreneurship and generation of further start-ups.

3.3.2 Venture Capital Description

The typical form of financing the development of start-ups is intended for venture capital. Historically, the first financing organization with the characteristics of modern VC was American Research and Development (ARD), formed in 1946 by MIT President Karl Compton, Harvard Business School Professor Georges F. Doriot, and local business leaders [29]. A previous similar case in financing of new technologies is that of Battelle Development Corporation (BDC), a not-for-profit subsidiary of the Battelle Memorial Institute, created in 1935 for financing internal R&D projects of Battelle Columbus Laboratories through their high-risk period to the point when industry will take them over [2], subsidiary already cited for the development of photocopy technology. Financing of technology developments by VC is radically different from that of industrial financing through R&D projects. The objective of industrial financing is to obtain an ROI by exploiting the new developed technology. The VC objective is completely different and consists of obtaining an exit by selling the developed technology or the entire business of a start-up and reinvesting the obtained capital from this ROI in new technology developments through start-ups in the form of cyclic activity. Another form of VC is that of business angels, individuals that invest their own capital in start-ups. Business angels do not normally have great capital as do large VC companies and tend to finance the initial phases of start-up development by sometimes taking more risky ventures. VC companies may be described in terms of structure, financing strategies, and selection methods in financing.

3.3.2.1 Structure of VC system

In territories where the presence of start-ups is important, the VC tends to be differentiated by one side following the innovation field of start-ups, for example, technology or socio-economics, while, on the other side, following the phases of development. In this case, there is trading activity among VC companies that are involved in the various phases of financing the start-up development, adapting in a certain way the increase of capitalization with the decrease of risk and increase of potential ROI. In this way, a market of start-ups is formed, with a list of capitalization values that are reported in local data banks in the Silicon Valley and are, of course, associated with high volatility. It should be noted that in the Silicon Valley, many big companies, initially created as start-ups, tend to continue to operate internally like a start-up and also to use their own large availability of capital to finance start-ups in competition with VC. In fact, big companies of this type prefer to make investments in start-ups, instead of in acquisitions.

There is also another type of competition with VC by team people who obtain large capital in the exit of their start-up and, in fact, also use these capitals to finance new startups. A historical case of this type was that of Steve Jobs, who invested in his own start-up NeXT and in the company Pixar with the obtained capital excluded from Apple [18].

3.3.2.2 Strategy of VC financing

The strategy of VC financing is conditioned by the achievement of a sufficiently high ROI in order to cover, not only the cost of development of the successful technology, but also capital invested in abandoned start-ups, and with possible margins for an increase in available capital for further investments. Such constraint limits the stages of start-up development in which the VC considers the investment [33]. In particular, a VC never invests in scientific research and almost never invests in proving scientific principle. It rarely invests in developing enabling technologies, but often invests in developing the use of a new technology or in developing a new product. Very often, a VC invests in revising and improving a product, broadening a product line, and applying a product to another application. Typically, VC considers various elements of risk in investing, such as size of the market and suitability of technology or product to the market's needs. It also considers the organization's building, the development plan, and people who should implement the plan, as well as the possibility of financing and the realistic method of exit from the investment [33]. In a general way, portfolio strategies, largely used in financial activities, are of limited value in a VC, not only in mitigating the downside risks inherent in science-based innovation projects, but also in enhancing the probability of exceptional rewards. VC portfolios must be closely scrutinized and cannot consider a high number of start-ups, such as that required in a portfolio strategy, in order to be successful. Consequently there is the unwillingness of the VC to invest in very young firms that only require small capital infusions avoiding the increase in the number of startups that must be scrutinized. The risk of technical failure involves achieving required performance specifications or matching market needs [33]. The alternative to portfolio strategy is to build a system of innovation designed primarily to maximize the probability of success for each project, limiting risk by selection of people in whom VC have confidence, in both the technical and the business aspects of the enterprise [35]. Looking at characteristics of territories with high VC activity, such as the Silicon Valley, a large number of financed start-ups is observed, accompanied by a high percentage of abandonment and very high ROI in cases of success; it seems that, even if each VC company does not follow a true portfolio strategy, the total set of VC companies in fact act following nearly a portfolio strategy with a high enough number of start-ups to statistically obtain a certain number of very successful investments. In fact, there are also particular strategies that exploit portfolio statistical advantages by making small investments in the feasibility of a very high number of start-ups, followed by the choice of a few of them that show the more favorable conditions for further investments.

3.3.2.3 Crowdfunding

Crowdfunding is a possible alternative to VC in financing start-ups that has found an important development in recent years. The capital of crowdfunding is collected typically through a campaign of funding with specific purposes and generally uses a dedicated internet platform, with funds coming more from individuals than from entities. Actors financing crowdfunding are thus principally a high number of individuals who participate with relatively small amounts of capital, but the participation with higher funding of other entities such as private or public organizations, banks, and firms is also possible. In fact, crowdfunding is a general business model that includes various types of financing, is mainly constituted by lending, followed by equity, and to a minor extent, by donations and even by do-it-yourself crowdfunding for personal purposes. Crowdfunding of start-ups is, in fact, a minor part of equity funding and has found great interest in countries where VC has low availability of capital and lower propensity to take risks. Startup crowdfunding finances typically the initial phases of development as seed capital, but in certain cases, it may also finance further increases in startup capital, and it has been observed that individuals tend to use their options in order to conserve participation in the development of a start-up.

3.3.3 The Startup – Venture Capital Cycle

The activity of the SVC system may be seen in the form of a cycle, as reported in Figure 3.4. The cycle starts with presentation of start-up projects to a VC that selects start-ups that can be financed by rejecting the others. Start-ups are financed and, following development, are abandoned or reach an exit. The technology or the entire business of start-ups is sold, and the obtained ROI is, in part retained by the VC, and the rest reinvested in new start-up developments starting a new cycle. In fact, part of an ROI might be also obtained in certain cases by founders of the start-up who sometimes reinvest in new startups, but we have not considered this possibility in our simplified view of the cycle. This cycle is valid for either a single VC or a group of VC of a territory, and it is characterized by a certain number of average selection rates concerning: rate of accepted startups for financing, rate of abandoned startups, rate of obtained ROI in relation to the total invested capital, and rate of part of ROI retained by the VC to cover its expenditures and rewards. At the end of the cycle, the difference between the obtained ROI and the total investment in start-ups, reduced by the amount of ROI retained by the VC, represents the possible surplus capital for reinvestment. Its value may be positive, or even negative, if the ROI is not high enough to cover the invested capital. The condition of financing equilibrium of the cycle is reached when the amount of obtained ROI, less the amount of return retained as margin by the VC, is equal to the initial amount of invested capital of the cycle. On the basis of these conditions of equilibrium, it is possible to develop a simple simulation model and to conduct parametric studies on the various conditions leading to reinvestments of available capital in new start-ups. This simulation model is

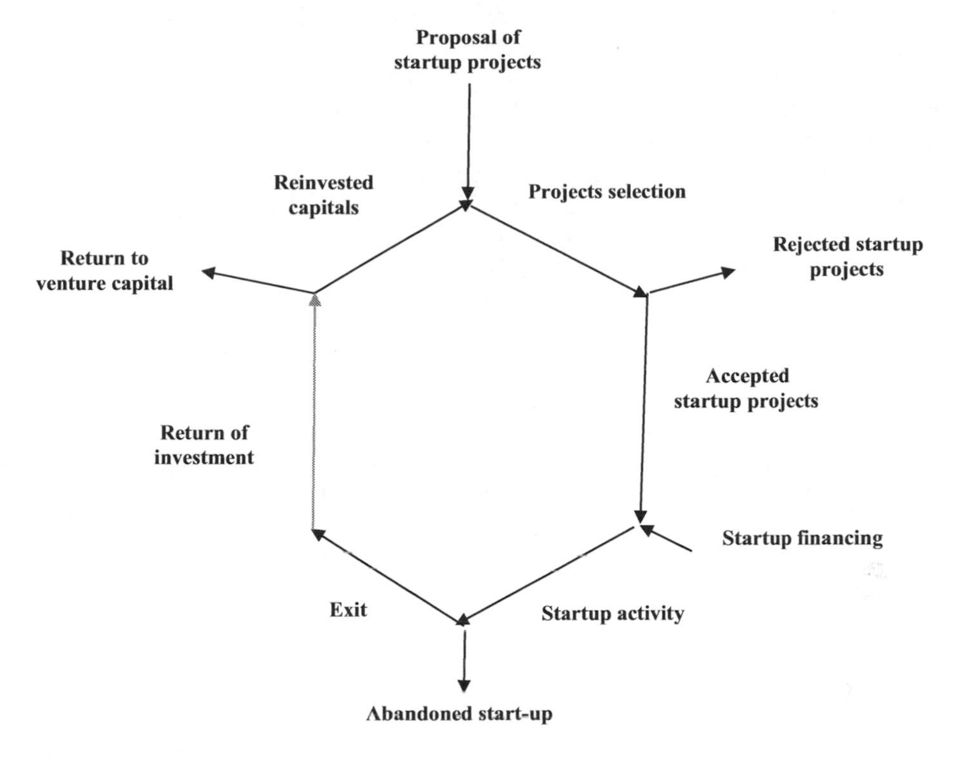

FIGURE 3.4 The SVC Financing Cycle.

mathematically described with its calculations in Appendix 3 at the end of this book and we give the main results here. For the parametric studies, the simulation model considers two different investing strategies called:

Strategy A. In this case, for selection, the potential return of investment of the start-up and the capacity and experience of the team are primarily considered; while feasibility is, of course, taken into account, but is considered too uncertain to be the main criterion of choice.

Strategy B. In this case, the feasibility of the start-up is of main importance in choice, and suitable methods for its determination are taken in consideration in order to reduce losses due to abandoned start-ups.

In a first run of calculations, we have considered the complete financing of the valid projects of startups and a rate of abandoning of financed start-ups of 90% for Strategy A. In the case of Strategy B, we have considered a variable acceptance of financed start-ups from 25% to 75% and an abandonment rate of 75%. In fact, these two strategies correspond, in the case of Strategy A, to a typical American strategy, while Strategy B is typically European. In the United States, a selection based on the potential generation of ROI and management quality of the start-up team would prevail. On the contrary, in Europe, the estimation of the chance of success of the start-up seems to prevail, while often the failure of a team in previous start-ups is considered negatively for further financial proposals. This attitude of considering previous start-up failure negatively is irrational; in this way, experience in management of previous start-ups is lost considering that, in every case, most start-ups are statistically abandoned. In fact, in the Silicon Valley, the success of a start-up is often obtained after two or even three failures on the part of a team. It should be noted, however, that presently, the European VC is increasingly adopting a type A strategy, accumulating experience, while the American VC is becoming more and more interested in also investing in European start-ups. The results of the model calculations show, as expected, the necessity of having a sensible higher value of ROI in the case of Strategy A in order to reach conditions of financial equilibrium and in fact, considering the adopted parametric values for the model, it is possible to calculate that the ROI of equilibrium for Strategy A should be 2.5 times higher than that of Strategy B and that this rate is dependent on the rates of abandonment of start-ups in the two strategies, but is, as expected, independent of the rate of acceptance for financing. Actually, the higher values of ROI obtained in Strategy A consequently also have a higher amount of available capital to invest in new start-ups, and this strategy should be considered more efficient from the economical point of view. Furthermore, it is shown than this advantage increases with the number of cycles considered in the simulation, and the model calculations, taking account of a sequence of cycles, also show that the ratio between financed start-ups and those corresponding to the financial equilibrium remains the same as in the first cycle for either Strategy A or Strategy B. Thus, the key factor making Strategy A more successful is not actually a consequence of the cycle, that in fact presents linear dependences following the adopted parameters rates, but by the fact that this strategy plays with a higher number of financed start-ups, selected on the basis of their high ROI potential and increasing the statistical probability of obtaining exits with high ROI. Following the SVC model simulation, we reach the conclusion that the high rate of accepted start-up projects and the high availability of financing capitals, joined with an effective selection of start-ups with a high ROI potential for financing, are at the base of success for Strategy A. This success is certainly also linked to the availability of an efficient know-how in selection and monitoring of the financed start-ups with a high ROI potential.

3.3.4 Difference between Industrial R&D System and SVC System

The SVC and industrially financed R&D projects systems present a certain number of similarities and some important differences. Comparing the schematic view of SVC and R&D activities reported in Figures 3.1 and 3.4, respectively, we may note that both present a financial cycle that, in the R&D process, concerns investment in developing new technologies, industrial capital for their use, ROI, available capital for investing in new R&D projects. In addition, in Figure 3.1 the R&D activity presents a knowledge cycle consisting of generating innovative ideas and presenting R&D project proposals, partly rejected and partly financed, that constitute the projects of R&D activity. In fact, there is also in the SVC system a knowledge cycle analogous to that of R&D that concerns the generation of an innovative idea for a start-up, its elaboration in form of a start-up project, presentation to VC for financing, selection of financed start-ups followed by start-up activities. As in the case of R&D, start-up activities generate knowledge by both success or abandonment that, as is the case for RDK in R&D activity [10], constitute a driving force for further start-up projects. The difference is in the fact that knowledge generated by start-up activities is enriched by nontechnical knowledge because start-ups carry out, not only R&D activities, but also business models and commercial developments. Considering now the financial cycle, there are actually radical differences in strategies between industrial capital and VC. In fact, industrial capital finances development of new technologies through R&D activities and looks for ROI based mainly on the use of new technologies. New investments in R&D by industrial capital do not depend on obtained ROI but on other factors such as economic situations, firm strategies, and existence of public aids. On the contrary, VC obtains ROI by selling the new technology just after its development and invests a part of the positive returns in the development of new start-ups, retaining only a part for their operative costs and revenues, as described previously, in the SVC cycle. It may be then observed that both R&D and SVC systems are characterized by the existence of autocatalytic processes of development, which are based in R&D on an increased return of knowledge by R&D activity that increases the number of R&D project proposals. In SVC, however, there is an increase in availability of capital for financing start-ups because of increased returns in capital obtained with exits. In these conditions, it may be observed that in the case of R&D the limits to development depends essentially on availability of investments for R&D, and in the case pf SVC, on the availability of valid start-up projects. The development of business models suitable to the developed technology exists in the SVC system but not in R&D projects activity; it may be observed that often companies are biased when making R&D investments in technologies that do not fit with their established business models, while start-ups have the advantage of freedom to establish the best business model suitable for the developed technology [35]. In fact, it is a common opinion that established firms exhibit a systematic biased underinvestment or overinvestment in commercialization of novel emerging technologies, while start-ups exhibit less of this bias. In other words, the underinvestment or overinvestment bias originates from considering an established model that may be not appropriate to the opportunities inherent in the new technology, while successful start-ups may better interpret the potential value of nascent technologies [35]. In industrially financed R&D projects, questions about markets and strategies are taken into full consideration only in the advanced phases of development and are mainly assumed by the firm financing the project. Consequently, only models of business compatible with the strategies of the firm are taken into consideration and, unlike the case of VC, a viable potentially successful new technology development may be abandoned if it cannot be included in the strategies of the firm [35]. Furthermore, the frequently observed superior efficiency of start-ups in the R&D process, in respect to industrial R&D process, apparently reflects the superior quality of their technical personnel, greater cost consciousness, better understanding of the problem to be solved resulting from closer contact with the firm's operations, and better communication [35]. On the other hand, the major limit of the SVC system is the need for development of new technologies with very high ROI in order to sustain and develop the financing of the SVC cycle. In these conditions, the cycle should be sustained by innovations, possibly with a high radical degree, but at the same time, have a higher risk of failures, as has been discussed previously. Another consequence of existence of the SVC system may be observed in big company strategies that limit R&D activity to their core business acquiring more radical technologies through start-ups [36]. Consequently. the main advantage of industrial financed R&D projects is the possibility to develop technological innovations with relatively lower ROI, eventually with a lower radical degree as well as lower risk of failure. The strategical situation for both R&D and SVC systems may be represented by the quadrants reported in Figure 3.5 that consider the possible choice of these two systems as a function of the degree of radicality of the developed technology and the expected ROI. The quadrant with high ROI and radicality is typically of interest in the SVC system, while the quadrant with low ROI and low radicality is typically of interest in the R&D system. The quadrant with high ROI and low radicality is, of course, of interest for both SVC and R&S systems although occurring much less frequently, while the quadrant with low ROI and high radicality is not of interest for both because of low ROI and higher risks of failure and consequent probable financial loss. Considering the previous discussions, a discussion remains open on the case of financing the development of small business projects or technologies of public interest, for example, in the environmental field, that are neither of interest for VC nor, often,

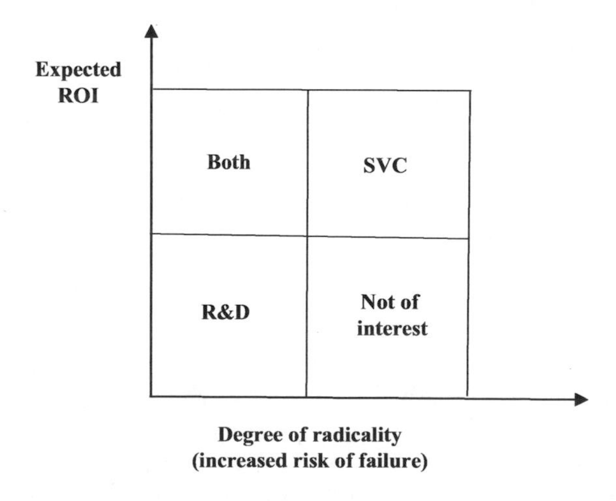

FIGURE 3.5 Fields Interesting SVC and R&D Innovation Systems.

for industrial capitals with no interest in niche markets. A problem of this type might also become important, for example, in SMEs of industrial districts for fostering the radical degree of their innovations and for implementing ICT in their manufacturing technologies. The possibility of using public financing of start-ups in compensating an insufficient availability of VC might generally be considered negatively because of lack of experience of public officials in sustaining a successful SVC cycle. In fact, it should be considered that the relation of VC to the team managing a start-up is not of simple supply and monitoring of the invested capital, but also of supply of contacts and strategies based on VC experience that is useful for the success of the start-up. However, public aid may be useful in the preliminary phases of development of start-ups looking for VC financing. The question of public US governmental financing of start-ups has been discussed, for example, in a workshop about managing technical risks [29] concluding in a series of recommendations for carrying out a public VC program including: building relationships and understanding of VC industry, considering the narrow technological focus and uneven levels of VC investments, appreciating the need for flexibility in the VC investment process, carefully analyzing the track record of entrepreneurs, and examining the track record of the firms receiving public venture awards [29]. Actually, the problem of public financing of start-ups remains in how to obtain enough ROI to sustain the financed SVC cycle without losses. Public financing may prevent retained revenues and operational costs, taken charge of by general expenditures of the administration. That may reduce the value of the equilibrium rate of the cycle and necessary ROI, but which may not be necessarily enough for financial equilibrium of the cycle. The open question is then: how much should public funds be continuously fed to the SVC cycle to maintain, in every case, a benefic general effect resulting from the development of start-ups?

3.4 THE INDUSTRIAL PLATFORM SYSTEM

The innovation of the industrial platform system, differently from industrial R&D and SVC systems, is not based on fluxes of knowledge and capital, but rather on relations between offers and demands of new technologies. These relations are different from the current discontinuous form that exists in the distributed technology innovation system, and are characterized by a continuous relation between the platform and the industry that requires technological advances and is not limited to the supply of technologies, but is also based on an exchange of knowledge useful for improvements and development of new technologies. An industrial platform may be seen as a group of firms managed by the owner of the platform that organizes fluxes of knowledge, technologies, and financial transactions. Industrial platforms are a particular type of general platform system of relations between demand and offer of products or services that has been developed especially in the social and financial field using ICT. It originated in the Silicon Valley and often exists in large firms typically offering to consumers, products or services such as Google, Apple, Microsoft, and Amazon, or specific services such as Uber, Airbnb, through great social networks such as Facebook or LinkedIn. Platforms dedicated to industry, i.e. industrial platforms, are only at the beginning of their diffusion, offering especially enabling technologies such as robotics, additive manufacturing, digital manufacturing, internet of things, big data and cloud computing, cybersecurity, and general software and hardware for artificial intelligence. Such offer is not largely diffused as in R&D or SVC systems. For these reasons, an important industrial experience that allows the development of quantitative simulation models and knowledge of real advantages and limits of this organizational structure (such as in the case R&D and SVC systems) is not available. However, technology dynamics may suggest some aspects that may characterize the effect of diffusion of such structures for industrial technological advances. For this purpose, it is useful to first present some general characteristics of platforms followed by a description of the peculiarities of the industrial platforms.

3.4.1 General Characteristics of a Platform

The concept of platform may be applied to any system of relations [9] and may assume various forms such as:

- Platforms for aggregations based on connections and transactions between users and resources
- Platforms based on social interactions and connections between individuals or communities
- Platforms based on mobilization and engagement toward long terms objectives
- Platforms facilitating learning and helping the improvement of individual capacities

Industrial platforms are a form of aggregation., however, like any other type of platform, they may also include, to a greater or lesser extent, the other forms of platform. The great development of the platform model may be attributed to the offering of major services around four major aspects such as: instant search and accessibility, possible direct intervention in creating custom solutions, fulfilling needs contextually when they occur, and interaction with people as human beings. The technology-driven effects of a platform are a self-reinforcing effect on competition that drives technology towards componentization (standardization of supplies), and searching more value and revenue within an Innovate-Leverage-Componentize (ILC) cycle in the presence of available commodities (internet, big data, cloud computing, etc.) [9]. The ILC cycle consists of a phase in which the system experiences some novelty (innovation) that gains, in a second phase, a leverage that generates demand for lower value components and transforms the novelty in a component (componentization). Platforms are the media for creation of a new environment generating totally new circumstances of work and value production that involve formerly considered consuming customers as peers in peer-to-peer systems of workflows and business models. According to these new possibilities, the nature of the firm itself is changing through a deeply disruptive process in the hierarchical management structure. Disruption is coming from transformation of *value chains* into multidimensional *value networks*. In the past, companies owned the different enablers and components of the business process, while nowadays they must excel at providing consumable interfaces and infrastructures [9]. Following the description of the various aspects of a platform, it is possible to distinguish various elements composing a platform and its structure [9]. The basic elements of a platform are owners, partners, peer producers, peer consumers, and stakeholders, described as follows:

Owners: These are represented by the proprietors of the platform, typically firms of various dimensions and even start-ups, but they may also include not-for-profit organizations, foundations, associations, cooperative structures. The owners assure the vision of the platform and its existence. Owners of industrial platforms are typically large industries in the field of hardware and software, but they may also be important consulting firms and even startups with a platform structure for specific technologies.

Partners: These are entities that have an intense and continuous cooperative relationship with the platform, creating its increased value. They provide useful complementary contributions to the platform with which they have agreements for a continuous collaboration.

Peer producers: These are entities that also provid useful and complementary contributions to the platform, but in a discontinuous way following the exigences of the platform. In industrial platforms, they may be also firms that are active in specific technological fields, hardware and software production, or technology consultants of interest for the platform.

Peer consumers: These are entities interested in buying the platform's products or services; in industrial platforms, they are typically firms that demand new technologies.

Stakeholders: These are entities interested in platforms and influencing their development or failure. Typically, they are public entities that regulate and control the platforms, with possible interest in their growth and prosperity. In the case of industrial platforms, they may also be public entities for promotion and aid in technological innovations for industry that include platforms.

A general structure of a platform may be seen as a series of concentric circles, as represented in Figure 3.6. At the center there are the owners, contiguous with the circle of owners there is the circle of partners. Externally there are the peer producers and more externally the peer consumers. Stakeholders are situated outside the circles as they are not directly involved in the activity of the platform. In this representation of the platform, the

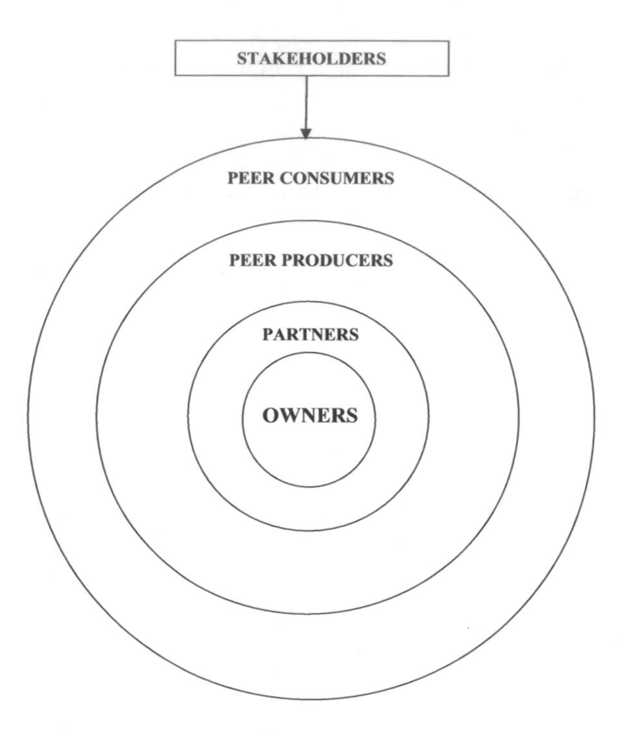

FIGURE 3.6 General Structure of a Platform.

position of circles represents the proximity of the actors of the platforms with respect to the owners, and the dimension of circles represents the number of actors of a certain type in the platform. There are also a certain number of processes occurring in a platform that may be described in terms of *transactions* and *channels* that in fact concern all the actors of the platform. Transactions may be monetary, for example, for an industrial platform covering the supply of products or services to peer consumer firms but may also be in the form of data or information that may be exchanged among the various actors of the platform. The channels are available structures facilitating the various types of transactions and have a crucial role in optimizing and maximizing the flux of transactions. Another important aspect of platforms involves the services offered internally to the platform and that are not assimilable to normal services supplied to peer consumers They are the *enabling services* that help the partners to generate value from their professional capabilities, for example, by gaining new markets, opportunities, and visibility, *strengthening services* that help peer producers in the transactions to improve their activities and possibly transforming them into partners. Finally, there are complementary services offered to the peer consumers in order to increase the value of the transaction, and then, the supply of additional products or services [9].

3.4.2 General Characteristics of an Industrial Platform

The industrial platform differs, with respect to R&D projects or start-up innovation activities, in the continuous availability of new technologies, not in discontinuous single events of technology supply that need further technology updating and leave finding solutions of specific problems associated with the supplied technology to the customer (peer consumer). This means that in an industrial platform, there is a much more developed flux of information and knowledge among actors developing or using the technology, favoring its updating but also the generation of ideas for the development of new technologies. In Figure 3.7 we report a schematic structure of an industrial platform from the technological point of view, and it should be noted that in industrial platforms, partners and peer producers may also be suppliers of nontechnological services, such as marketing and commercial services. The owners have relations and monetary transactions with partners, peer producers, and consumers, and informative relations with stakeholders. In particular, the owners have monetary transaction with peer consumers who pay for the obtained technological products and services, but between them an exchange of knowledge in two directions also exists that is useful to improve and update supplied products and services and for the development of new technologies. In the case of partners and peer producers, the monetary transaction occurs essentially from the owners toward these two platform actors that supply technological products and services to the platform, the difference being that partners have a continuous relation, while peer producers have a discontinuous

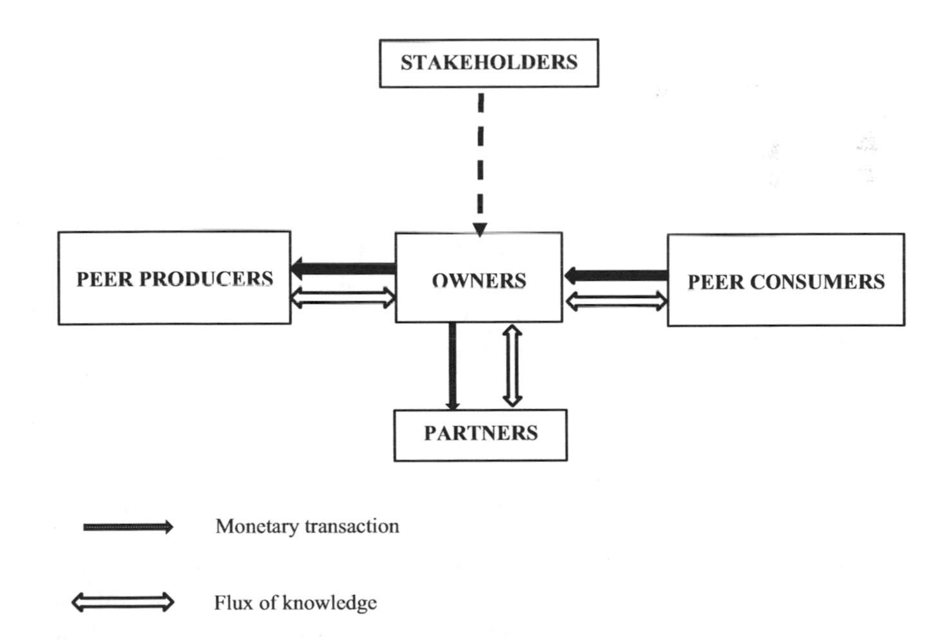

FIGURE 3.7 Technological Structure of an Industrial Platform.

relation that depends on the exigence of the platform. Both partners and peer producers have, as do peer consumers, an exchange of knowledge in two directions that is useful to improve and update supplied products and services and for the development of new technologies. Monetary transactions such as payments, and exchange of information of a technological nature are carried out through suitable channels, systems of payment, and methods for exchange of information that favorize maximization and optimization of their fluxes. As in the case of general platforms, there are enabling services in industrial platforms that help the partners and peer producers to generate value from technological innovations and in a possible transformation of peer producers to partners. Peer consumers also receive complementary services of a technological nature concerning the possibility of increasing the value of supplied products and services. There is, of course, an important difference between industrial platforms and general platforms operating in social, commercial, and economic fields. This difference concerns how an offer of a technological nature may be considered by a peer consumer, i.e. a firm, needing technological advances. In fact, the offer of a technology for industrial purposes needs solidity and reliability to be accepted, which cannot be simply the result of a rapid reaction and instantaneous acceptance as may occur in commercial or social platforms, neither can acceptance be influenced by a number of *likes* as they occur on the social platforms. In fact, the acceptance of an industrial product or service may need a period of evaluation that delays decisions, differently from general platforms, which are characterized by instant search, accessibility, and decisions about acceptance [37]. On the other hand, in the industrial platform, although a typical platform for aggregations, there are also such possible effects as formation of a community, mobilizations toward a common objective for innovation, and improvement of capacities for all the actors in the industrial platform. The introduction of industrial platforms in the innovative system of a territory may cause a certain number of changes in transfer of technologies and competition because of certain advantages offered by platforms. In fact, industrial platforms are not a single firm actor, but a variable group of firms that evolve from the generative ideas occurring in the platform and from the needs of either platform firms that offer technologies or peer consumer firms. Platforms thus enter into competition with suppliers of R&D services such as contract research organizations, public or private laboratories, and universities, which could shift offers to single firms toward offers to platforms. Furthermore, start-ups that have an available technology may look for an exit within a platform instead of with a company. On the other hand, competition among firms offering new technologies might shift toward competition among platforms. The great exchange of data and knowledge between the platform and peer consumers about the implementation of a new technology may have a benefic effect on the necessary learning curve for the use of the technology, because the platform can make a lot of knowledge coming from various peer consumers available that may be useful to reduce the learning time to rapidly reach optimal conditions for using the technology.

3.4.3 Suitability of Various Technological Fields for Industrial Platforms

There is a question about suitability of industrial platforms for the various technological fields. It is well known that platforms are well suited to ICT, having the advantage of a continuous innovative relation with peer consumers with respect to the industrial R&D or SVC systems of innovation. However, the possible success of industrial platforms is not known in the case of emergent new technological fields such as synthetic biology, nanotechnologies, and green technologies. In fact, there are no reasons that hinder the use of industrial platforms with their advantages in any technological fields, but what should be considered as a limit is the loss of freedom by peer consumers to make a choice of alternative technologies by other sources without interrupting the continuous relation with the platform with its possible disadvantages. Then a firm may consequently be motivated to remain linked to a platform with advantages to obtain continuous technological innovations. However, remaining in a continuous exclusive relation with a platform may have also disadvantages if, as a result, technologies of the platform become less competitive and the process of innovation becomes too slow or less effective with respect to other sources of new technologies. Actually, in ICT it seems that specific reasons exist to link peer consumers at a platform such as that observed, for example, in the supply of a specific informatic operative system that links the firm and peer consumer, to applications (technologies) for various purposes possible in the frame of the used operative system that is a property of the platform. If the firm would be interested in a technology of another platform it would need the operative system and technology of the other platform, and it is consequently, also faced with the necessity of a possible change of all applications (technologies) linked to the previous operative system; this may be an obstacle to the adoption of a technology of another platform. It is interesting to examine whether this type of platform advantage also exists in technological sectors other than ICT.

3.4.3.1 Synthetic biology

Among the various biotechnological applications that synthetic biology represents, it is probably the technology more similar to ICT from the point of view of suitability to a platform system of innovation. In fact, this similarity to ICT is based on the informatic molecular nature constituted by the polymeric DNA, the giant molecule able to generate specific languages for the construction of proteins, and possibly derived products, with a great number of potential applications. Basically, a synthetic biological process starts with the design of a DNA sequence, suitable for the desired purpose, that should be attached to the DNA of a microbiological entity, called chassis, free of the part of the original DNA that is not necessary for its existence. In fact, the formed DNA corresponds to a computer memory containing programs that may be operated, and pieces of this DNA constitute a part that encodes a biological function (protein coding sequence). Other necessary molecular parts that should be present are promoters, operators, ribosome binding sites, and terminators that have the correspondent functions of initiating the transcription of the DNA part, regulating the transcription, initiating the translation encoding the protein in the ribosome, and ending the transcription. These molecular parts constitute devices that may be assembled analogously to devices composing an electronic circuit, and there are analogous devices constituting biological logic gates or having signaling or controlling functions. The assembling of various suitable devices enables the synthesis of the desired product such as a protein from the designed DNA, and possibly, other molecular products derived from proteins. The difference from electronic circuits is that devices are not connected physically but contained in the cellular liquid in which are realized the wanted processes for an application [38]. With the development of this technological sector, future formation of platforms would be possible in which the owners and their partners control a proprietary core process of formation of suitable DNA structures by introduction of specific DNA parts in a microbiological entity for various specific production purposes, while peer producers may be involved in the development of specific molecular circuits necessary to operate the proprietary assembling of DNA that can produce the desired proteins and possibly protein-derived products. In this case, the key technology is represented by the proprietary method of assembling the DNA of the owners, while the operative system composed of suitable biomolecular devices depends on the desired products, and is of interest to partners or peer producers. Peer consumers of this type of platform are represented by firms involved in specific types of products such as pharmaceutics and chemicals, as well as firms involved in, for example, biofuels, sensors production, synthetic tissues that are interested in using synthetic biology technologies for their productions. Systems of gene editing for assembling DNA are already known, but variants and alternative methods may be developed and patented, especially considering that synthetic biology may synthetize DNA containing bases differently from natural ones, and then possibly proprietary, giving proteins containing amino acids that are different from natural amino acids, also possibly proprietary, suitable for the various applications. Although this possibility of use of artificial components has been proven until now only in one case [39], it might open a further field of large possibilities for synthetic biology that are similar to ICT, although the preparation of molecular devices and designed assembling is much more difficult than in the case of the design and construction of electronic devices.

3.4.3.2 Nanotechnologies

These technologies do not have an informatic base like ICT and synthetic biology. Nanotechnologies are based on the exploitation of various different properties of matter when it is reduced to a nanometric dimension. That makes possible the exploitation of properties of nanoparticles, nanometric layers on surfaces, and building of complex nanometric objects and even of molecular machines. Applications may include advanced material production, as well as electronics, medical devices, pharmaceutical products, and surfaces with special properties. However, because of the various types of products and applications, it does not seem possible to have a technological core that is in measure to form a platform with possible peer producers and consumers in the same way as in the case of ICT and synthetic biology, with the corresponding advantages.

3.4.3.3 Green technologies

Green technologies are involved in many fields such as production of renewable energy, elimination of pollution, and environmental technologies limiting energy consumption and emission of pollutants. All these technologies are very diversified from the technological point of view and they do not seem suitable or have the advantages of industrial platforms for the same reasons cited for nanotechnologies.

3.5 POSSIBLE EVOLUTION OF THE VARIOUS ORGANIZATIONAL STRUCTURES

When discussing the three organizational structures and their possible evolution from the technology dynamics points of view, it is important to consider the evolutive character of these structures, starting with the birth of R&D activities in the second half of the nineteenth century, followed by the offer of R&D services for industry at the beginning of the twentieth century, the formation of a system of distributed innovation and an open innovation regime in the second half of the twentieth century that generated the development of new organizational structures such the SVC system and the industrial platform system at the beginning of the twenty-first century. In fact, the formation of these three important organizational structures represents the various possible solutions to the evolving differentiation of technologies following their higher or lower degree of radicality, and the various potentials in returns on investments, accompanied by the great importance assumed by science and technology in the modern society. As we have previously discussed, the differences between the industrial R&D projects system and the SVC system concern the radically different strategies in financing and exploiting the new technologies. These differences made the SVC system more suitable for more radical technologies accompanied by high returns of investment, while the R&D project system is more suitable for new technologies of a relatively incremental type with relatively low returns of investment. The industrial platform system is different, because based on a different system of relations, and not on the base of different financial strategies, returns of investment, or radicality of the developed technology. In fact, industrial platforms are a more complex system that include R&D project activities and possible presence of start-ups financed by VC. In a certain way the industrial platform system has superior characteristics due to the continuous supply of new technologies, occurring within an enlarged system of development of new technologies with a high degree of flexibility and adaptive behavior, particularly by the presence of peer producers. On the other side, there is in the industrial platform system a great exchange of knowledge and experience among, not only the producers, but also the users of a technology, making a more favorable environment for the improvement and development of new technologies. In fact, the large availability of generated RDK is considered in technology dynamics as a driving force for innovations in the R&D system, as well in the SVC system and in industrial platforms. Furthermore, technology dynamics has shown that the use of technology is an important source of new technologies and that it is an important aspect of industrial platforms in which there is an exchange of knowledge between the peer consumers using the technology and the platforms. In the future development of industrial platforms, these factors might have a key role in its diffusion, independently of absence of advantages discussed for ICT and their possible existence in synthetic biology applications as well. In conclusion, the existence of these three forms of organizational structures for innovation might evolve toward a complex process of technological innovation activities with the formation of a more complex system for innovations constituted by networks of industrial platforms, including both R&D and SVC activities, linking generators and users of new technologies and exploiting the advantages of cooperation in the frame of important fluxes of knowledge. The industrial platform system is only at the beginning of its development; we have discussed certain specific advantages that exist in platforms using ICT, but that do not seem be present in other technological fields. It is possible, however, to imagine a scenario in which the platform system becomes prevalent in the activity of technology innovation, disrupting the present technological distribution system based on R&D activity internal to firms, and discontinuous research contracts and cooperation leading to a system that is reported schematically in Figure 3.8. In this figure, the innovation system is composed of a firm (peer consumer) linked to a network of platforms, supplying technologies related to the different sectors needed by the firm. These platforms, constituted by owners and their partners, are in contact with various peer producers that might also cooperate discontinuously with different platforms and consist mainly of, for example, small companies, start-ups, independent research laboratories, contract research organizations, universities. It should be noted that the firm is indirectly also linked to other firms (peer consumers) in an indirect exchange of, for example, data and information, and RDK, through their linked platforms, and then indirectly to all involved partners and peer producers. In this system, the various types of industrial platforms might be also formed in technologies that do not have advantages of an informatic nature as reported for ICT and synthetic biology. For example, in the field of nanotechnologies, the formation of platforms specializing in nano-particle production or in nanometric surface layers would be possible, and in green technologies, platforms specializing in solar photovoltaic systems or batteries would be possible. In this way, a firm might enter into a relation with various types of platforms, thereby covering its entire technology innovation needs that not necessarily covered by internal R&D, or even putting into discussion its R&D activity in competition with platform services. In this case, platforms would act as concentrators of RDK and information for the development of new technologies, as well as of LbyD experience contrasting effects of externalities and intranalities that exist in the technologies of peer consumers. Furthermore, platforms might distribute information for the search of solutions in the network of peer producers that might actually also cooperate with many other platforms. This may present advantages because technology

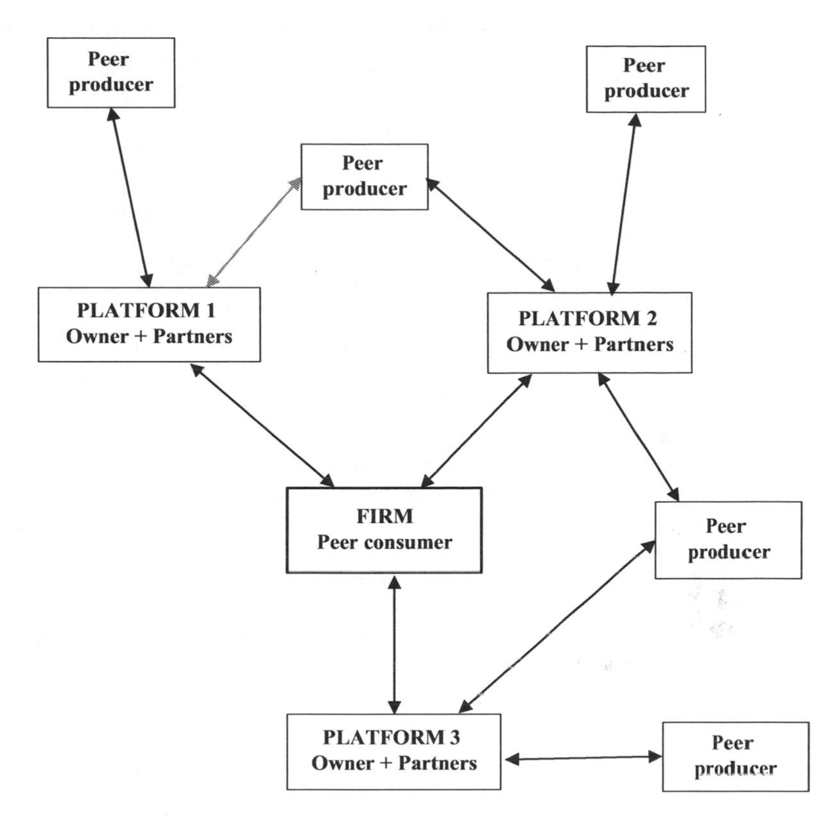

FIGURE 3.8 Industrial Platform Network.

dynamics has shown that new technologies may be formed from the ramifications of technologies developed for other purposes. The consequence will be an enormous generation of fluxes of data, with information and knowledge boosting technology improvements, and generation of ideas for new technologies constituting the possible strength of the industrial platform network system. The limits to the development of such systems of technology innovation are linked in a certain way to the attitude of firms to maintaining knowledge and practices secretly instead of exchanging knowledge, in order to obtain superior advantages. For this behavior, a characteristic Italian expression exists: *Pulcinella' secret.* Pulcinella is a Neapolitan theater mask that is believed to have an advantageous secret knowledge, but in fact, this knowledge is also available to other people who may also believe it to be their secret advantage. Such behavior is diffused in various industrial sectors and districts with respect to technological practices, leading to a system with a poor exchange of knowledge and RDK with stagnation and even technological decline. In fact, Pulcinella's secret with the Red Queen regime represent the major causes of technological stagnation and decline of industrial districts and sectors. The future will determine whether such behavior will be overcome, and whether industrial platforms will really take a great place in technology innovation because of the previous cited advantages transforming the innovation process into a complex system of generation of innovative ideas, financing, developments and relations.

REFERENCES

[1] Basalla G. 1988, *The evolution of technology*, Cambridge University Press, Cambridge UK.

[2] Boehm G., Groner A. 1972, *Science in the service of mankind, the Battelle story*, Lexington Books. D.C. Heath and Company, Lexington, KY.

[3] Rhodes R. 1986, *The making of the atomic bomb*, Simon & Schuster, New York.

[4] Feldman M.P., Francis J. 2002, The entrepreneurial spark: individual agents and the formation of innovative clusters, 195–212, in *Complexity and Industrial Cluster*, Quadrio Curzio A., Fortis M., Editors, Physica-Verlag, Heidelberg.

[5] Berreur L., Guillot P., Lesrel O. 1989, *Les Organismes de Recherche sous Contrat dans la CEE*, Commission of the European Communities, EUR 12112 FR-EN. Brussels.

[6] Malerba F. 2007, Innovation and the evolution of industries, 7–27, in *Innovation, Industrial Dynamics and Structural Transformation*, Cantner U., Malerba F., Editors, Springer Verlag, Berlin.

[7] Haour G. 2004, *Resolving the innovation paradox: Enhancing growth in technology company*, Palgrave Macmillan, St. Martin Press LLC, New York.

[8] Chesbrough H.W. 2003, *Open innovation: The new imperative for creating and profiting from technology*, Harvard Business School Press, Boston.

[9] Cicero S. 2017, *From Business Modeling to Platform Design*. https://platformdesigntoolkit.com/platform-design-whitepaper/.

[10] Bonomi A. 2017, A technological model of R&D process, and its implications with scientific research and socio-economic activities, *IRCrES Working Paper*, 2/2017.

[11] Dumbleton J.H. 1986, *Management of High Technology Research and Development*, Elsevier Science Publishers. New York.

[12] Thompson J.D. 1968, o*Organizations in Action*, McGraw-Hill. New York.

[13] Bonomi A., Marchisio M. 2016, Technology modelling and technology innovation: how a technology model may be useful in studying the innovation process, *IRCrES Working Paper*, 3/2016.

[14] Arthur B. 2009, *The nature of technology*, Free Press, Division of Simon & Schuster Inc. New York.

[15] Griliches Z. 1992, The search for R&D spillovers, *Scandinavian Journal of Economics*, 94 (Supplement), 29–47.

[16] Bayoumi T., Helpman E., Coe D.T. 1996, R&D spillover and global growth, *NBER Working Paper*, 5628.

[17] Saxenian A. 1994, *Regional advantage*, Harvard University Press. Cambridge, MA.

[18] Isaacson W. 2011, *Steve Jobs*, Simon & Schuster. New York.

[19] Scott D., Grindell A.G. 1967, *Components and systems development for Molten-Salt Breeder reactors*, ORNL – TM – 1855. Oak Ridge, TN.

[20] Fleming L., Sorenson O. 2004, Science as a map in technological search, *Strategic Management Journal*, 25, 909–928.

[21] Ben-David J. 1968, *La recherche fondamentale et les universités. Réflexions sur les disparités internationales*, OCDE, Paris.

[22] Bonomi A. 2014, Bridging organizations between university and industry: from science to contract research, *Working Paper Cnr-Ceris*, 15/2014.

[23] Lam A. 2011, What motivates academic scientists to engage in research commercialization: “gold”, “ribbon” or “puzzle”, *Research Policy*, 40 (10), 1354–1368.

[24] Gackstatter S., Kotzemir M., Meissner D. 2012, *Building an Innovation-driven Economy – The Case of BRIC and GCC Countries*, Basic Research Program at the National Research University, Higher School of Economics, Moscow, Russian Federation.

[25] Becker B. 2013, *The Determinants of R&D Investment: A Survey of the Empirical Research*, Economics Discussion Paper Series, School of Business Economics, University of Loughborough, Loughborough, UK.

[26] Bilbao-Osorio B., Rodriguez-Pose A. 2004, From R&D to innovation and economic growth in the EU, *Growth and Change*, 35 (4), 434–455.

[27] Haour G., Miéville L. 2011, *From science to business: How firms create value by partnering with universities*, Palgrave Macmillan. St. Martin Press LLC, New York.

[28] Bonomi A. 2017, A mathematical toy model of the R&D process: how this model may be useful in studying territorial development, *IRCrES Working Paper*, 6/2017.

[29] Lerner J. 2000, When bureaucrats meet entrepreneurs: the design of effective "public venture capital" programs, 80–91, in *Managing Technical Risk: Understanding Private Sector Decision Making on Early Stage Technology-Based Projects*, Branscombe L., Morse K., Roberts M., Boville D., Editors, NIST GCR 00-787, US Department of Commerce. Washington, DC.

[30] Bonomi A. 2019, The start-up venture capital innovation system: comparison with industrially financed R&D projects system, *IRCrES Working Paper*, 2/2019.

[31] Morris R. 2014, The first trillion-dollar startup, *Endeavour Insight Monthly Newsletter*, July 26, 2014.

[32] Chesbrough H.W., Rosenbloom R. 2000, The dual-edged role of the business model in leveraging corporate technology investments, 57–62, in *Managing Technical Risk: Understanding Private Sector Decision Making on Early Stage Technology-based Projects*, Branscombe L., Morse K., Roberts M., Boville D., Editors, NIST GCR 00-787, US Department of Commerce. Washington, DC.

[33] Morgenthaler D. 2000, Assessing technical risk, 104–108, in *Managing Technical Risk: Understanding Private Sector Decision Making on Early Stage Technology-Based Projects*, Branscombe L., Morse K., Roberts M., Boville D., Editors, NIST GCR 00-787, US Department of Commerce. Washington, DC.

[34] Scherer F.M. 1999, *New perspectives on economic growth and technological innovation*, Brooking Institution Press, Washington, DC.

[35] Morgenthaler D. 2000, More ways to fail than to succeed: strategies for managing risk, 61–120, in Branscombe L., Auerswald P. *Taking Technical Risks*, MIT Press, Cambridge, MA.

[36] Branscombe L., Morse K., Roberts M., Boville D. 2000, *Managing Technical Risk: Understanding Private Sector Decision Making on Early Stage Technology-Based Projects*, NIST GCR 00-787, US Department of Commerce. Washington, DC.

[37] Hobcraft P. 2018, The emerging world of connected industrial ecosystems, *Ecosystem4innovators*, January 10, 2018.

[38] Freemont P.S. et al., 2016, *Synthetic biology: A primer*, Imperial College Press, World Scientific Publishing Co. Singapore.

[39] Callaway E. 2017, Cells use "alien" DNA to produce protein, *Nature*, 551, 550–551.

4 Applications of Technology Dynamics

4.1 TECHNOLOGY DYNAMICS AND THE INNOVATION PROCESS

In Chapter 2, which focuses on technology, we provide a first perspective on the innovation process, considered as a sequence of steps from the generation of an innovative idea to the use of the new technology; that has been represented in Figure 2.7. Following the description of technology dynamics in terms of processes and structures of technology innovation, we provide a more articulated view of technology innovation, taking account of various possible paths and conditions, followed by an innovative idea from its generation to its transformation in a new technology and finally entering into use, describing, in a certain measure, the existing complexity of technology innovation. From the perspective of technology dynamics, technology innovation may be described by taking in consideration three main phases that concern: the generation of the innovative idea, the development of this idea until the formation of a new technology, and the further generation of technologies during the use of this technology. These phases are described by technology dynamics in the following sections.

4.1.1 Generation of Innovative Idea for the Technological Innovation Process

The generation of an innovative idea is fundamentally a combinatory process involving pre-existing technologies with or without exploitation of new (or never exploited) phenomena discovered by science. This generative process may be the result of individual creativity, or it may emerge through generative relations among various actors. The combinatory process normally also includes general scientific and technical knowledge, as well as other types of knowledge. An important exploitable source of knowledge results from the activities of organizational structures that lead to technology development and involve successful or abandoned projects or start-ups in the respective R&D and SVC systems, as well from relations in the industrial platform system that exchange knowledge among owners, partners, peer producers and peer consumers.

4.1.2 Development of the Innovative Idea to the Formation of a New Technology

This development may occur generally in one of the three different organizational structures: the R&D project system, the SVC system, or the industrial platform system. In these structures, there are different financing strategies for innovation, or different relations between offer and demand of new technologies. The use of a specific organizational structure for the development of new technologies tends to be differentiated based on the degree of radicality of the developed innovation and its potential return of investment, as well as by particularities of the technology that favor the industrial platform system.

4.1.3 Generation of Innovations during the Use of a Technology

During the use of a technology, there are externalities and intranalities that influence the efficiency of the technology, leading to the search for new, better conditions of operation or even making some changes in its structure and generating a new technology that is normally of an incremental type. In less frequent cases, a new idea is born, leading to the development of a more efficient, alternative radical technology. Furthermore, the demand for improvements, diversifications, and alternatives to the used technology may generate a ramification of technologies, typically of an incremental nature, such that, although each new technology is of limited application, their great number forms the bulk of the socio-economic impact of the original technology. Furthermore, by the spandrel effects, new radical technologies with the same or different purposes may appear to generate further ramifications that impact the evolution of technologies. Another important consideration during the use of a technology is the velocity of innovation existing in its technological field that might rapidly make available new, more efficient technologies and make a chosen technology already obsolete during its implementation. Finally, during the use of technologies in industrial districts or sectors producing the same types of products, formation of the Red Queen regime often occurs, characterized by district or sector firms developing only incremental innovations with technological, but not economic, development. Such regimes may be disrupted by the entrance of a radical technology of production that may eliminate firms of the district or sector that are unable to restore their technological efficiency.

4.2 TECHNOLOGY DYNAMICS AND STATISTICAL STUDIES

Statistical studies are largely used in investigations about technology innovations and its economic effects. Technology dynamics is not based on statistical studies but on description of processes and structures of technology innovation and it should not be considered an alternative to statistical studies, but rather a complement that suggests the limitations of current statistical studies, which are guided, for example, by rules of the Frascati [1] or Oslo [2] manuals to explain technology developments. The contribution of technology dynamics consists of explaining the origin of collected data and then the dynamics of technology innovation behind its measured effects in the socio-economic system. On the other hand, technology dynamics might suggest possible new types of statistical studies based on the detailed aspects of the studied innovation processes and structures. In order to explain, through the technology dynamics vision, the limited possibilities of studies of innovation with current statistical data collections, we have considered three types of studies: the first concerning the general study of research and innovation activities in territories, the second concerning the relation between investments in R&D and economic growth, and the third concerning the study of patents in two cases: 1) relation of patents to scientific research and 2) patents values distribution.

4.2.1 Research and Innovation Studies

Research and innovation activities are commonly considered important for the economic growth of a country and these activities have an important role in governmental policies for economic development. For many years, research and innovation activities have been objects, especially in OECD countries, of statistical studies with the aim of giving support to innovation policies. For this purpose, OECD countries have developed a guide to standardization of data collection since 1964, to compare statistical data of various countries. This guide, the *Frascati Manual*, has been updated until publication of the last available edition in 2015 [1]. More recently, another guide, the *Oslo Manual* has been introduced [2], which generalizes the concept of innovation beyond the R&D activities considered by the *Frascati Manual* and provides a larger field for the collection and recording of data. The *Oslo Manual* considers R&D to be just one of the various identified innovation activities. This point might be controversial as, in fact, R&D activity, and in general, technology innovation, may be considered the direct or indirect sources of actual innovation activities through ICT, artificial intelligence, and possibly, in future synthetic biology applications. These technology innovations would possibly determine the existence and shaping of many new social and economic innovations in sectors such as financial technologies, and social networks that involve, for example, software innovations or actual innovation management cited for statistical studies in the *Oslo Manual*. For these reasons, the fundamental aspects of technology development, possibly at the origin of further types of innovations, should be taken into special consideration with respect to the other types of innovations for the establishment of policies that could potentially contribute to the economic growth and prosperity of a country. In order to understand the possible contribution of technology dynamics to research and innovation studies, it is important to consider the history of the definitions of R&D activities that are actually reported in the *Frascati* and *Oslo* manuals and used in guiding measurements of scientific, technological, and innovation activities. In fact, the origin of their definition of R&D activity may be attributed to a report by Christopher Freeman, Director of the Science Policy Research Unit of University of Sussex, in a draft document prepared in the autumn of 1962, and revised and accepted later by OECD countries [3]. This view has evolved into a more accurate definition, as reported in the last edition of the *Frascati Manual* [1]. Following this manual, the definition of R&D comprises: a creative and systematic work undertaken in order to increase knowledge, including knowledge of humankind, culture, and society to devise new applications of available knowledge. Furthermore, the manual considers R&D composed of three types of activity: basic research, applied research, and experimental development. Such activities are not considered in a temporal sequence because of the various directions of knowledge flow, and their definitions have been previously presented in Chapter 2, in a detailed discussion of the stages of the innovation process. On the basis of these definitions the manual presents classifications of institutional sectors, provides specific guidance in measuring R&D for the various sectors carrying out this activity and examines effects of globalization of R&D on statistical studies. Technology dynamics, unlike the *Frascati* and *Oslo* manuals, neatly separates scientific research from R&D while observing the existence of intertwining processes between them. This separation has been already anticipated in the discussion of the steps of the innovation process and it is coherent with the definition adopted by technology dynamics on the nature of technological innovation, which is seen as a combinatory process of pre-existent technologies that either exploit or do not exploit scientific discoveries. This separation is fully justified in the frame of studying the innovation process from the generation of an innovative idea to its transformation in a new technology. In fact, scientific research was not taken into consideration in earlier definition and management of R&D activities that developed in the United States since the 1930s. This view, combined with the establishment of rules for a best practice and coherent with the reality of this activity, may be attributed to Clyde Williams, Director and later President of the Battelle Memorial Institute from 1934 to 1953. Differently from Cristofer Freeman, he was not an academic figure but a chemical engineer who previously worked at the US Bureau of Mines and faced the problem of making a contract research activity with industry economically viable. He introduced some basic rules for best practice in R&D activity with respect to an entrepreneurial attitude for researchers, management by project objective in an organizational matrix structure, importance of economic aspects of R&D projects, and autonomous R&D financing; this anticipated, to a certain extent, the foundation of a subsidiary, the Battelle Development Corporation, in 1935, which was typical of the SVC system concept [4]. These basic management rules have been largely accepted in the United States and are also presently followed in the Silicon Valley. Actually, Battelle has been appreciated in the United States more for its capabilities in R&D management than for stimulating industrial research [5]. Separation of R&D activities from scientific research has been also adopted in the Dumbleton's handbook, previously cited in the discussion of the technology innovation steps, that considered the OECD model of R&D unnecessary when discussing R&D activity with the objective of development of new technologies [6]. It is interesting to note that this separation has been in fact not a limitation, but an important stimulus, to develop scientific research with the aim of exploiting its results, and in the case of fundamental research, to develop of radical competitive new technologies with a high ROI potential. From the American point of view, there is not really a difference between oriented and basic research, as defined by *Frascati Manual*, because both are considered important, being soon or later, the possible origin of radical innovations with high competitiveness. In conclusion, technology dynamics does not consider pure basic research or fundamental research as part of R&D, but considers oriented research as the scientific part of an intertwining process with R&D. Applied research and experimental development described in the *Frascati Manual* are considered in technology dynamics as steps in the development and industrialization of technologies. Technology dynamics finally highlights the importance of efficiency in exploiting knowledge as an important factor that determines the results of research and innovation activity in terms of economic growth and prosperity in a country. All these considerations do not mean, of course, that the statistical studies about the various types of innovations guided by the *Oslo* or *Frascati* manuals are losing importance. In fact, they can provide a detailed view of the research and innovation activities in a country, which is certainly useful for contributing to the establishment of policies that potentially help economic growth and prosperity in a country. Technology dynamics may simply pose some limits to the *Frascati* and *Oslo* manuals as they do not take into account the evolution and integration of R&D in more complex organizational structures, particularly SVC and the nascent industrial platform systems, with their different impacts on technology innovation activities and effects on the socio-economic system. In fact, data collected based on the cited guides have referred more to the effects of innovations in the socio-economic system than to processes and structures generating such data. From the scientific point of view, data collected on effects in the socio-economic system would represent the kinetics, rather than the dynamics, of technology innovations. Technology dynamics might then suggest further statistical studies of the actual processes and structures of technology innovation in order to also elaborate methods measuring the efficiency, and not only the extent of innovation activities. In fact, innovation should not be only considered in terms of quantity of input, considering R&D investments, structures, and personnel, but should also be considered in terms of effectiveness and output amount; for example, R&D investments only constitute a rough measure of the level of innovation of a firm or of a country [7]. Such measures of efficiency might be, in certain cases, more useful than measures of activity in the elaboration of policies for promotion of innovation in a territory. However, it should be also noted that technology dynamics may simply give indications for new types of statistical studies related to the studied processes and structures, but it cannot verify their feasibility or elaborate methods for collecting and reporting the data, which could be a task for statistical guiding manuals. Finally, considering the case of globalization of R&D activity, technology dynamics observes that, if future globalization becomes largely diffused, activities of generation and exploitation of new technologies might be dispersed in various countries and might change continuously as a function of strategic decisions of the companies. In fact, globalization can geographically separate the function of the generation of innovative ideas from the function of developing the correspondent new technology, as well from the function of use of the new technology for production and finally the taxation of ROI generated by technology exploitation. In this case, the rules of Frascati or Oslo manuals in creating comparable innovation activity data for various countries, would lose validity because of a continuous variable dispersion of the innovation activities in various countries, with loss of the connections existing in a country between the generation of an innovative idea, the development, and the exploitation of the new technology. New radical approaches in determining the existing comparable innovation activities in a country would be necessary.

4.2.2 R&D Investments and Growth

The relation between R&D investments and national or regional economic growth is a major subject of statistical and econometric studies. Technology dynamics may lead to a certain number of observations about assumptions and discussion of results of certain aspects of these studies. In particular, there are two aspects questioning the relation that exists between R&D investments and economic growth. The first aspect concerns the fact that the R&D model of technology dynamics does not consider any decrease in rate of generation of new technologies, or growth when there are enough available investments for valid R&D projects, contrarily to many results of econometric studies reviewed recently in which an inversed U behavior has been often observed regarding the effect of various determinants on R&D [8]. The second aspect concerns the importance given by the model to the efficiency of the innovative systems of territories, instead of simple amount of R&D investments, in the relation between R&D investments and growth, because of the presence of different conditions of distributed and open innovation activities. These differences lead to varying availability of RDK, also due to the existence of R&D that is not related to economic purposes and differences in exploitation efficiency of scientific results, as discussed previously in the comparison of American and European visions of entrepreneurial or cultural use of scientific research, and finally also possibly to the existence or nonexistence of technological innovations that are independent of research activities. Concerning relation between R&D investments and growth, econometric and statistical studies in which it has been observed an inverted U curve, the following study is cited as an example [9]:

> In this study it has been discussed the dependence in various countries of Gross Domestic Product (GDP) per capita, expressed as average of period 1998–2001, on Gross Domestic Expenditure on R&D (GERD), expressed as percentage of GDP, as average of period 1996–1997, then assuming a time lag between R&D investments and resulting GDP of about 2–4 years, using World Bank data, finding an inverted U curve (quadratic function). This study shows also that values of GDP of various industrialized countries, following the quadratic curve obtained by the econometric model, tends to increase less than proportionally to the increase of GERD and even to show a decrease as in the case of Sweden, observing, however, certain deviations with higher values of GDP in the case of USA and Italy and a lower ones in the case of Japan in respect to the calculated curve. The observed decrease of GDP with the increase of GERD was attributed to diminishing returns to research investments that play a similar role to diminishing returns to capital accumulation into standard neoclassical growth model. This study shows also that

a level of GERD percent equal to 2.7 maximises the GDP per capita, value close to the suggested objective of GERD of 3% for the European countries by the Lisbon strategy of EU [10].

As already cited, the R&D model of technology dynamics cannot explain a decrease of GDP with an increase of GERD and, on the contrary, it foresees an exponential increase in new technologies, and then in growth with the increase of R&D investments. On the other hand, the pertinence of the standard neoclassical growth model as explanation might be doubtful considering that technology is not a true economic good and that the neoclassical model would not be valid in studying effects of new technologies on the economic change [11]. Following the previous considerations, it seems that the observed inverted U behavior does not actually have a satisfactory explanation. Technology dynamics might explain this incongruity by the fact that growth does not depend only by R&D investments, but also on the efficiency of the innovation system of the territory. A striking example of probable quasi inexistence of limits to economic growth associated with technological developments is represented by the Silicon Valley. Its innovative system has been studied in the frame of the winning competition with the electronic industry of Route 128, a region near Boston [12]. In this territory, firm strategies are based on continuous development of innovations, organizing activities in term of projects and subcontracting production, indifference to cannibalization of own products," products, and certainty that this strategy will necessarily generate economic growth. Innovation activity in the Silicon Valley is not seen in terms of R&D investments, but as a general capital investment that includes industrialization and commercialization of products, as well as venture capital financing start-ups indiscriminately for both R&D and business model developments. In fact, Silicon Valley, with the long term technical and economic success of its innovative system appears as an important challenge to traditional economic views on relations between investments in R&D and growth. Regarding agreement of data with the curve obtained in the considered study [9], we observe quite scattered positions for countries with low GERD, a good agreement in the case of France, Germany, the United, Kingdom and Sweden, but sensible deviations, as cited previously, in the case of Italy, the United States and Japan; the first two have a value of GDP much higher, and the last one a value of GDP much lower, with respect to the curve. The dispersed data for GDP for low GERD value may be easily explained by the importance of other factors in the building of GDP with respect to investments in R&D activity in types of countries such as those previously observed, for example, in the case of countries that export or do not export oil [13]. For the important deviations observed for the United States and Italy, we might advance an explanation considering the differences in the innovative systems of these two countries with those of the other European countries. In fact, the American innovative system is characterized by a large availability of RDK, mainly through military R&D, which increases births of new technologies and reduces costs of R&D for their development and better exploitation of scientific results, as previously discussed. On the contrary, in Italy, the deviation might be explained by a real value of GERD that is much higher than the recorded value, due to an innovative system linked to the existence of large number of SMEs, often organized in industrial districts, that do not necessarily take account of the cost of their innovation activity in terms of R&D investment, observed in previous studies [14]. Another concern about the study is that it represents a nearly static situation interpreting a period of only five years, and, in fact, there are no available changes in GDP with time as a function of large changes of GERD values for single industrialized countries. As a consequence, Italy, for example, should double its GERD value to reach the suggested optimal value by the Lisbon strategy; these optimal value is apparently reasonable. However, there are no prior historical real examples confirming that an important increase in GERD has the direct effect of a large increase in GDP in industrialized countries. Actually, there is an interesting example in the case of South Korea of important economic growth with an important growth of R&D investments at the same time, but in which growth cannot be considered as simply the result of a high increase in R&D investment alone [5]:

> In the sixties South Korea was an underdeveloped country got out of a war with R&D investments lower than 0.5% of its GDP. In 1965 president Park negotiated with USA a loan of 150 million US$ with scientific and technical assistance to develop scientific research and R&D in the country. The aid included the realization of two main research centres: the Korean Advanced Institute of Science (KAIS), with the help of Frederik Terman the godfather of the Silicon Valley, the Korean Institute for Science and Technology (KIST) for contract research organized by Battelle, later merged with KAIS becoming the Korean Advanced Institute of Science and Technology (KAIST). Such investments, accompanied by suitable policies, were successful generating an important economic growth and R&D investments reaching values around 3% of GDP in the nineties. However, examination in detail of the followed policies it appears that success of South Korea cannot be attributed only to technological innovation efforts, and then R&D investments, based on model existing in USA, but also by adopting Japanese industrial organization [5].

The history of South Korea developments shows that simple increases of R&D investments are not sufficient to assure economic growth. On the contrary, the possibility might be considered that the observed high increase of R&D investments is, in fact the result, and not the cause, of its successful model of research, radically changed by the creation of KAIS and KIST joined with boosting technical education and suitable industrial organization in this developing country. This may suggest that the South Korean experience may hold important lessons for both developed and developing countries [5].

4.2.3 Patent Studies

Patents are practically the only direct source of public information about the activity of technology innovation and, for this reason, are frequently considered by statistical studies concerning innovative aspects or resulting contributions to economic growth, as well as the relation of technology innovation to scientific research. In Chapter 2 ("Technology"), we present a definition of patents and their relation to the model of technology and, in order to discuss the contribution of technology dynamics to patents statistical studies, we chose to consider two examples of studies. The first study concerns a better understanding of the mechanisms connecting scientific research and exploitation of knowledge by measuring the time lag that exists between scientific research and patent publications, and that in the case of nanotechnologies and of polymer technologies [15]. The second study concerns distribution of patent values and their implication in technology policy [16].

4.2.3.1 Relation between scientific research and patents

In the first patent study [15], there is an interesting tentativeness for a better understanding of the mechanisms connecting scientific research and exploitation of knowledge in order to provide important information to policy-makers. The idea is to measure the time lag existing between scientific research and patent publications, assuming that the faster scientific literature is cited in a patent, the more the technology sector is knowledge intensive. In this context, both the nanotechnology sector and the polymeric materials sector have been compared. The study shows that in nanotechnology the time needed for scientific knowledge to be incorporated into a technical application is around 3 – 4 years, confirming the existence of a high intensity knowledge in this field. This is not observed in the polymer field in which dependence is much flatter without an important significative maximum that might indicate a specific time lag between scientific research and patents. Technology dynamics might suggest a further, more articulated explanation for this difference. Nanotechnologies are formed by quite varying sectors of applications for nanoparticles, nanometric surfaces, nanometric composites, and even molecular machines that have been studied relatively recently. This means that nanotechnology ramifications are not very developed and are still quite dependent on scientific research and R&D that explain the clear links between scientific publications and patents. Polymer technology, on the contrary, results from much older scientific research, and is expected to present a developed technology ramification and then be composed of corresponding patents derived more from the use of this technology than from R&D and scientific research, which explains the difference between the practical absence of connection between the two technologies. In conclusion, the author also admits the impossibility of going further by investigating whether patents are exploited or not exploited, and then makes an estimation of the real economic impact of patents related to scientific research; this argument is discussed subsequently, considering the process of generation of patents from scientific research that is based on the results of technology dynamics studies.

4.2.3.2 Patent values distribution

The second patent study [16] draws evidence regarding the size distribution of returns from eight sets of patents that are attributable to private sector firms and universities. The distributions are in most cases highly skewed, and it was observed that the top 10% of patents sample captured from 48 to 93% of total returns values in all the samples. The greatest considered patent sample included 772 German patents that were filed in 1977 and held valid until their expiration in 1997. In this sample, only five were able to generate very large patent returns. In fact, 81% of total estimated value was observed in the top 10% of patents, and a similar value of 81% – 85% was observed for a sample of 222 US patents. Even higher values between 91% and 93% were observed in the case of patent royalties of six universities. Lower percentages, around 60%, were observed in the case of start-ups and initial public stock offerings, and even lower values around 48 – 55% were observed in the case of two estimations for pharmaceutical products patents. We may observe that this study considers indicatively that a representative sample for statistical studies is in the order at least of several hundred patents. Actually, this study on patent values represents a good example of methodological approach to statistical patent evaluation, but it also shows how a long period of time should be considered to obtain reliable evaluation of the success of patents. In fact, the study indicates that there is a difference in the distribution of patent values obtained after a stabilization time of use of a technology, and the results shows that results are less skewed in the case of start-ups and IPOs patent values. On the other hand, a special case is observed of pharmaceutical industry patents with even a lower skewness. These deviations indicate a lower skew distribution of returns for IPO, and pharmaceutical products might be explained by technology dynamics, considering that there is a previous higher selection with respect to generally delivered patents. in the acceptance for IPO for new technologies, while in the case of pharmaceutical products there is a selection because of the various protocols to be respected in order to accept the product for efficient treatment of human disease. This preliminary selection might explain the lower skewness of returns by elimination of patents that have low returns or are less efficient. Finally, the skew number of patents with high value, with respect to the total number of patents protected during their period of validity, observed in many studied cases, might indicate that many patents are protected during their validity more for defensive reasons than for ROI, but also that many patents result from ramifications of radical technologies, very numerous but with limited applications, and then with limited ROI.

4.2.3.3 Generation of patents from scientific research

Technology dynamics may easily explain the limits of patents studies in determination of the impact of technologies on the socio-economic system because of the complex processes in which scientific results are involved in the formation of successful patents that contribute to the economic growth of a country. This may be appreciated, considering the patents steps sequence consisting of scientific research, generation of innovative ideas, patent applications, delivery of patents and, finally, their conservation of rights and use as reported in Figure 4.1. The process by which innovative ideas coming from scientific research result in

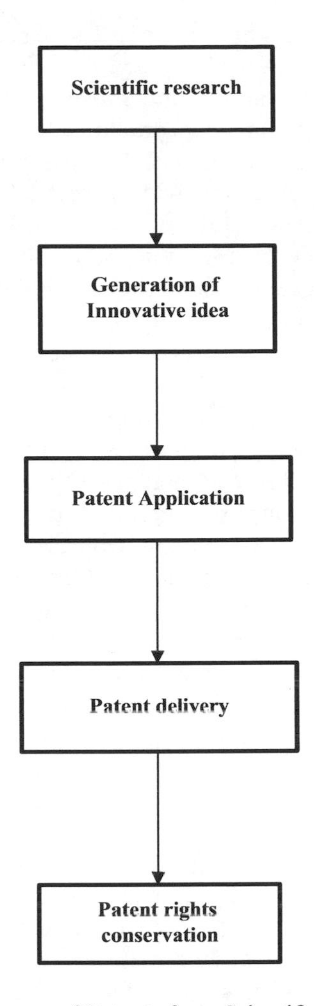

FIGURE 4.1 Generation Process of Patents from Scientific Research.

delivered patents is highly selective, and an important part of the generated R&D projects are abandoned and many patents applications are rejected. Furthermore, patents that originated directly from scientific research are, in fact, a minority with respect to patents originated by ramification of technologies and with respect to improvements and diversifications obtained during the use of technologies. These last-mentioned types of patents are neither directly related to any scientific result, nor in some cases to any R&D activity, but to simply exploiting general scientific knowledge for technology improvements [17]. On the other hand, it may be observed that the economic success of a patent does not necessarily depend on its scientific origin, nor on the amount of investment in technology development. In fact, most of the patents are applied in R&D after the feasibility step, and not in the industrialization step when the development of the technology is nearly terminated. Nevertheless, a scientific origin favors the radical degree of the innovation and consequently, a possible future technology ramification process with its impact. It should be also noted that not all delivered patents are publicly available, but may be covered by secrecy for military reasons, as was the case for the Szilard invention of nuclear weapons discussed previously, with respect to the interface between R&D and scientific research, but nevertheless later contributing to the advancement of technologies and growth. On the other hand, there is also the case of used technologies that are not patented and prefer the secrecy of the invention and know-how, considering the public delivery of a patent as a danger for unwanted imitations of the technology. In conclusion, while scientific activity may be estimated relatively easily on the basis of publications and citations, the obtention of reliable knowledge on technology innovations from statistical study of patents is much more difficult. In fact, in addition to the previous considerations, many delivered patents are not exploited, others have their protection abandoned after a few years; even recently some patents have been obtained by nonpracticing entities that are merely trying remuneration on discoveries (made by themselves or third parties) from operative companies that have interest in these patents [18]. Technology dynamics may suggest a different approach to the study of the impact of technology development in specific territories in which patent studies are not considered. This may be done by considering not only the amount of research and innovation activity and derived patent activity, but also by a benchmarking evaluation based on the innovation processes and structures present in the territory. For benchmarking, it is possible to use the best practices for technology innovations, reported, for example, in a previous study [19] in which the Silicon Valley has been considered as a useful territory for benchmarking. Such an approach might establish, differently from patent studies, not only potential, but also chances, in terms of probability of successful technology developments in discussing similarities and differences referring to the territory used for benchmarking. Actually, it should be noted that the benchmarking study would not simply duplicate a reference system, such as the Silicon Valley system, in favor of another territory. In fact, a tentative transfer of the Silicon Valley system to the East Coast of the United States failed, even with the help of Frederick Terman, the godfather of the Silicon Valley [5]. What is possible is not a transfer of the entire system to a new territory, but rather, considering the various elements composing the benchmarking system studying modification and recombination of these elements in such a way that matches both the different environments and the possible improvement of the innovative efficiency of the studied territory.

4.3 TECHNOLOGY DYNAMICS AND PROMOTION OF INNOVATION

A first consideration derived from technology dynamics studies about promotion of innovation is that the frequent suggested policy consisting of a simple increase of availability of capital for financing technology innovation does not necessarily constitute an effective promotion of innovation. Following technology dynamics, investments and financial aid should be considered as just a means that should be available to support the really efficient actions of promotion that consist of supporting the entire process of innovation in the frame of a valid strategy of use of organizational structures operating in a valid industrial organization. In fact, from this point of view, an intervention that is able to improve the efficiency of the innovation process might even give positive effects, not as the result of an increase in financing, but by being the cause of the increase in financing. In order to consider all the various types of actions for the promotion of technology innovation from the technology dynamics point of view, it is useful to take account of the various steps in the innovation process from the generation of the innovative idea to the use of the technology, as reported in Figure 2.7. Certain types of actions will, of course, differ, based on the organizational structure in which the innovation is developed, particularly the different cases of the R&D project system and the SVC system:

4.3.1 Actions for the Generation of Innovative Ideas

This step is favored particularly by creativity and entrepreneurial views about results of scientific research and available RDK. The important climate and organizational aspects that favor contract research and spin-offs from research laboratories for the creation of start-ups and the favorable environment for the generation of innovative ideas have been previously described in the discussion in Chapter 2 about the stages of the technology innovation process. Generative actions are helped by technology transfer offices of universities or other promoting organizations. A particular case of promotion of innovative ideas concerns the case of industrial districts that make the same products, in which it is possible to establish cooperation for R&D activities and generative relations for innovative ideas. This leads to the emergence of innovative R&D projects useful for the solution of common problems. A case of this type has been described in the frame of Italian districts for the production of valves and faucets [20] and it is subsequently described in detail with respect to specific promotion of technology innovations in SMEs.

4.3.2 Actions for the Feasibility Step

The importance of carrying out this step concerns the fact that its results are critical for obtaining either industrial or venture capital funds for the development of innovative ideas. This suggests the necessity for laboratories to have available funds for preliminary research about the innovation and possibly for carrying out short studies on its economy and markets in order to attract financing of their activity. In the case of start-ups there is the need for seed capitals that might also be in the form of public aids or crowdfunding.

4.3.3 Actions for the Development and Industrialization Steps

The key action of this step concerns, of course, the availability of industrial financing for R&D projects or venture financing in the case of start-ups. In these steps, the frame in which the activities are carried out is important, and best practices suggest an organization of R&D that is project- oriented and includes a matrix structure in which project leaders have great autonomy in the frame of a hierarchical structure with only administrative tasks and of settling of conflicts.

4.3.4 Actions during the Use of the Technology

The use of technology is considered by technology dynamics as an important source of further new technologies through a ramification process that it is favored by the existence of technicians with a good intermediate scientific and technical education. Another aspect, not often considered is the generation of new technologies, is the simple combinatory nature of generation of technology that is favored by a large knowledge of available technologies, not necessarily of the same technological sector, and may be diffused, for example, by competence centers or other structures that may be particularly available to SMEs of industrial districts.

4.3.5 Promotion of Technology Innovation in SMEs

In conclusion, it is necessary to note that SMEs have different problems in the promotion of technology innovations than do large firms. Large firms normally have enough available capital to invest in R&D projects but that is not often the case for SMEs. In fact, costs of R&D projects for the development of a technology are largely independent of the size of the investing firm and the dimension of market covered by related new products or services. Many SMEs cover only niches of market and cannot foresee possibly very high ROI. For this reason, technology innovations of interest for SMEs are generally not of interest for VC, but availability of R&D investments are essential for maintaining the technological competitivity of the firms, and then, of industrial districts. A possible solution practiced in industrial districts that make the same type of product is the formation of a cooperation network with a certain number of firms providing common knowledge and financing in order to be able to carry out R&D projects with a common use of the results. It is important in the case of industrial districts that this network might include not only producing firms, but also subcontractors currently involved in certain operations of production in order to have availability of all competencies of the production chain, thereby avoiding problems of intranality effects. A case of cooperation of this type that led to the formation of a consortium for R&D activity is Ruvaris, described as a successful case of cooperation for innovative technologies [20]. It has this interesting history of cooperation of SMEs in industrial districts:

> Ruvaris was born as effect of a multiclient study launched by a technological park located in the north-east of the Piedmont region of Italy in 1996. It was based methodologically on the experience of Battelle in multiclient studies for great companies that was transferred and adapted to the reality of SMEs active in production of valves and faucets of two industrial districts in the provinces of Novara and Brescia. The study concerned surface treatment and new materials technologies that might be considered by this industry and obtained the participation of 25 firms. The project was carried out in 1997 and, at the end of the study, six firms decided to form a company, called Ruvaris, to develop a technology for the elimination of lead on the surface of brass in order to comply new rules about contamination of drinking water. The R&D project was concluded successfully in 1999 and technology patented in USA and Europe, the first industrial plant entering in function in 2000 followed by diffusion of the technology in Italy and abroad reaching the number of 15 plants in 2005. After a period dedicated to the certification of the process, in 2005 Ruvaris launched a new multiclient study about future lines of innovation obtaining the participation of 19 firms. At the end of the study it was decided to form a consortium for R&D activities including projects carried out in partners laboratories and in cooperation with universities. This consortium is presently active with the participation of about 20 firms of the sector of valves and faucets production.

Ruvaris is a good example of cooperation on R&D among SMEs, enabling the solution of many problems concerning development of new technologies in industrial districts.

4.3.6 Possible Evolutions of Cooperation in Production Technologies

Another challenging evolution in the SME manufacturing industry includes the introduction of ICT, robotization, and internet of things in production. This transformation may require high levels of investments that are largely independent of the size of production of SMEs that might be not able to have the necessary capital to upgrade their manufacturing technology. In the case of industrial districts producing the same type of product, it would be possible to consider, not only cooperation in R&D activities, such as in Consortium Ruvaris, but possibly in a common production plant fully robotized and using ICT and flexible enough to be capable of manufacturing all types of products of the participating firms. Such a plant might be a solution for the availability of enough capital for the necessary technological transformations by subdivision of the total among the various partners and covering the total productivity of the plant at the same time. In this case, an industrial district that already has common subcontracting firms for certain technological operations, would have also a common production plant for the final product, while the various firms would have activities devoted essentially to the design of the product and its commercialization. Such transformation, joined with common R&D activities, could made an industrial district nearly equivalent to a large industry producing the same product but with the advantage of having a richer variety of products and a large exchange of knowledge due to the existence of the network among numerous firms.

REFERENCES

[1] OECD. 2015, *Frascati Manual 2015: Guidelines for Collecting and Reporting Data on Research and Experimental Development*, The Measurement of Scientific, Technological and Innovation Activities, OECD Publishing, Paris.

[2] OECD/Eurostat. 2018, *Oslo Manual 2018: Guidelines for Collecting, Reporting and Using Data on Innovation, 4th Edition*, The Measurement of Scientific, Technological and Innovation Activities, OECD Publishing, Paris, Eurostat, and Luxembourg.

[3] OECD. 1963, *Proposed Standard Practice for Surveys of Research and Development*, Directorate for Scientific Affairs, DAS/PD/62.47, Paris.

[4] Boehm G., Groner A. 1972, *Science in the service of mankind, the Battelle story*, Lexington Books. D.C. Heath and Company, Lexington, KY.

[5] Stuart W., Leslie S.W., Kargon R. 1996, Selling Silicon Valley: Frederick Terman's model for regional advantage, *The Business History Review*, 70 (4) (Winter 1996) 435–472.

[6] Dumbleton J.H. 1986, *Management of high technology research and development*, Elsevier Science Publisher, New York.

[7] Haour G. 2019, The crucial human factor in innovation, 393–412, in *The Routledge Companion to Innovation Management*, Chen J., Brem A., Viadot E., Wong P.H., Editors, Routledge Taylor & Francis Group.

[8] Becker B. 2013, *The Determinants of R&D Investment: A Survey of the Empirical Research*, Economics Discussion Paper Series, School of Business Economics, University of Loughborough UK.

[9] Coccia M. 2008, Science, funding and economic growth: analysis and science policy implications, *World Review of Science, Technology and Sustainable Development*, 5 (1), 1–27.

[10] Room G. 2005, *The European challenge: Innovation, policy learning and social cohesion in the new knowledge economy*, The Policy Press, Bristol.

[11] Nelson R.R., Winter S.G. 1982, *An evolutionary theory of economic change*, Belknap Press of Harvard University Press, Cambridge MA and London.

[12] Saxenian A. 1994, *Regional advantage*, Harvard University Press, Cambridge, MA.

[13] Gackstatter S., Kotzemir M., Meissner D. 2012, *Building an Innovation-driven Economy – The Case of BRIC and GCC Countries*, Basic Research Program at the National Research University, Higher School of Economics, Moscow, Russian Federation.

[14] Hall B.H., Lotti F., Mairesse J. 2009, Innovation and productivity in SMEs: empirical evidence for Italy, *Small Business Economy*, 33, 13–33.

[15] Finardi U. 2011, Time relations between scientific production and patenting of knowledge: the case of nanotechnologies, *Scientometrics*, 89 (1), 35–50.

[16] Scherer F.M., Haroff D. 2000, Technology policy for a world of skew-distributed outcomes, *Research Policy*, 29 (4–5), 559–566.

[17] Fleming L., Sorenson O. 2004, Science as a map in technological search, *Strategic Management Journal*, 25, 909–928.

[18] Haour G., Miéville L. 2011, *From science to business*, Palgrave Macmillan, St. Martin Press LLC, New York.

[19] Bonomi A. 2018, Sistemi Innovativi Tecnologici Territoriali. Due casi: il Verbano Cusio Ossola e il Canton Ticino, *Quaderni IRCrES*, 1/2018.

[20] Rolfo S., Bonomi A. 2014, Coopération pour l'innovation au niveau local: un exemple italien de succès, *Innovations*, 44, 57–77.

5 Conclusions

The description of technology dynamics in this book supports the idea that technology innovation is a nonlinear process characterized by the existence of critical levels of activity above which are technological developments and below which are technological stagnation and decline. Many innovation processes may be described through triggered cycles generating autocatalytic technological developments or slackening evolutions that terminate with the arrest of generation of innovations. This vision of technology innovation differs somewhat from some common views regarding the origin of new technologies, how technologies influence economic growth, and what are the adoptable policies for promotion of innovations. A first difference concerns the possibility of generating important technologies by a simple combinatory process of pre-existent technologies, without direct exploitation of new or never-used phenomena discovered by science, and nevertheless leading to important innovations. The possibility of generating new technologies by a simple combinatory process makes possible other ways of generating innovations that do not necessarily depend on typical R&D activities and on exploitation of scientific results, but rather on technical work carried out especially during the use of technologies. Although many such innovations may appear to be of limited importance, their very high number is responsible for the development of technology ramifications that, in fact, are at the origin of the bulk of the socio-economic impact of an original radical technology. Contrary to a common opinion, the use of technologies, not only research, is therefore an important source of new technologies. The adopted model of technology defines technology innovation as the result of changes in the structure of a technology, of either a radical or simply incremental nature, resulting from activities not only by R&D as well as by LbyD. Both activities may be seen in the technological landscape as a search for an optimal recipe for the technology, possibly also innovating by changing the structure of the technology forming a new landscape. Technology dynamics considers important the various types of organizational structures in which are developed new technologies such as R&D, SVC and the industrial platform systems. Such structures are the result of an evolution of the way in which technology innovation. is carried out and they might evolve in the future to form a complex network system for generation of innovative ideas, financing, development, and relations. Another crucial aspect discussed by technology dynamics concerns the importance of the technological innovative efficiency of a territory, defined as the efficiency of exploitation of available knowledge, not only from scientific research, but also from previous R&D activities in the frame of what is called a distributed system of innovation in a regime of open innovation. Furthermore, it should be considered that available knowledge for innovation does not come only from research activities with economic purposes, but also from research conducted for military purposes, space exploration, and sometimes, even from equipment developed for scientific purposes. In conclusion, technology dynamics considers that economic growth and prosperity of a country does not actually depend on R&D investments, which should rather be considered a means, but on the intensity of generation of innovative ideas, which depends on the efficiency of exploitation of knowledge by the territory, and of course, by adopted strategies and availability of capital for financing technology developments in the frame of a suitable industrial organization. The articulated views of technology dynamics regarding technology innovation lead to considerations that effective policies for promotion of innovation may be established not only by increasing research and innovation activities, but also by improving the efficiency that characterizes these activities, and improvements that may be obtained by considering the most suitable organizational structures and the best known innovation practices.

Appendix 1
Mathematical Model of Technology

A1.1 TECHNOLOGY MODEL

This mathematical model considers a technology as a set of technological operations that may be structured as a graph with the corresponding matrix. Each operation is characterized by a certain number of instructions or parameters and each parameter may assume a discrete number of values or choices in a certain range of variability. For example, a heat treatment technology may be composed of three operations: heating, temperature maintenance, and cooling. Heating is characterized by parameters, such as heating velocity and temperature, that should be reached; maintenance is characterized by maintaining time and temperature; and cooling is characterized by cooling velocity. Each parameter may assume a certain number of values within a certain range. Technology may thus be described as a structure of operations represented by an oriented graph in which nodes represent the starting/ending points of an operation and arcs the operations. This graph is similar to representation of tasks used by the PERT method in project management. A simple example of oriented graph structure for the heating technology constituted by three arcs in sequence and their associated parameters has been presented in Chapter 2 ("Technology") in Fig. 2.2.

Following this model, a technology may be defined by a set O composed of N operations:

$$O = \{o_i, \ i = 1, \ldots, N\} \tag{1}$$

Each operation o_i is characterized by a set M_i of M_i specific instructions or parameters:

$$M_i = \{p_{ij}, \ i = 1, \ldots, N; j = 1, \ldots, M_i\} \tag{2}$$

in which p_{ij} represents the jth instruction associated with the ith operation o_i. The total number P of instructions characterising a technology is given by:

$$P = \sum_{i=1}^{N} M_i \tag{3}$$

The instruction p_{ij} may assume a set S_{ij} of different values or choices:

$$S_{ij} = \{s_{jik}, i = 1, \ldots, N; j = 1, \ldots, M_i; k = 1, \ldots, S_{ij}\} \quad (4)$$

in which S_{ij} indicates the cardinality of the set S_{ij}.

The N operations cannot be considered simply a set; in fact they normally have a specific temporal sequence that may be represented by an oriented graph. Indicating with ***E*** the set of events determining the start or/and end of the operations and, as previously, with O the set of the operations, we can build up a graph τ that we can call graph of the operations of the technology:

$$\tau = (E, O) \quad (5)$$

in which ***E*** represents nodes and ***O*** represents the oriented arcs of the graph. In our model, we take into account that each operation can be associated with more than one instruction as in Equation (2). For example, an operation such as heating in a heat treatment can be associated with an instruction as the final temperature, but also with a specific velocity of heating. With N the number of operations being from Equation (1) and P the total number of instructions being from Equation (3), we have:

$$P \geq N \quad (6)$$

When N = P each operation is characterized by only one instruction or parameter.

A1.2 TECHNOLOGICAL RECIPES AND TECHNOLOGICAL SPACE

Considering a specific technology with a set of N operations corresponding to a total of P instructions, we can define as *technological recipe* the specific configuration ω obtained attributing a specific value or choice to each of the P instructions. The set Ω of all the possible configurations of a technology is given by:

$$\Omega = S_{11} \times S_{12} \times \ldots \times S_{1M1} \times \ldots \times S_{NMN} \quad (7)$$

In other terms we have:

$$\Omega = \{\omega_1, 1 = 1, \ldots, \prod_{i=1}^{N}\prod_{j=1}^{M_i} S_{ij}\} \quad (8)$$

The number of configurations $|\Omega|$ is given by:

$$|\Omega| = \prod_{i=1}^{N}\prod_{j=1}^{M_i} S_{ij} \quad (9)$$

Should $S_{ij} = S$, $i = 1, \ldots, N$ and $j = 1, \ldots, M_i$ we have:

$$|\Omega| = S^P \tag{10}$$

We may note that the number of configurations varies exponentially along with the number of values or choices for the instructions and even with a small number of instructions, the number of technological recipes is very high.

In order to better explain the previous equations, we may illustrate a simple example considering a technology with the number of operations N = 2 and then:

$$O = \{o_1,\ o_2\}$$

Should, for example, operation o_1 be a heating and operation o_2 be a cooling we have:

$$M_1 = \{p_{11}, p_{12}\}$$

Where the operation of heating is associated with $M_1 = 2$ instructions such as p_{11} as the final temperature and p_{12} as the velocity of heating, we may have, at the same time for the operation o_2 of cooling:

$$M_2 = \{p_{21}\}$$

that corresponds to a free cooling to a final temperature indicated by instruction p_{21}. Now considering that there are two possible heating temperatures and only one value of velocity of heating we have:

$$S_{11} = \{s_{111},\ s_{112}\};\ S_{11} = 2$$
$$S_{12} = \{s_{121}\};\ S_{12} = 1$$

At the same time, should two be the final cooling temperatures we have:

$$S_{21} = \{s_{211}, s_{212}\}\ ;\ S_{21} - 2$$

The number of configurations ω present in the set Ω would be four:

$$|\Omega| = S_{11} \times S_{12} \times S_{21} = 2 \times 1 \times 2 = 4$$

These configurations or technological recipes may be represented as:

$$\omega_1 = (s_{111}\ s_{121}\ s_{211})$$
$$\omega_2 = (s_{111}\ s_{121}\ s_{212})$$
$$\omega_3 = (s_{112}\ s_{121}\ s_{211})$$
$$\omega_4 = (s_{112}\ s_{121}\ s_{212})$$

We may also define a Hamming distance d among the recipes as the minimum number of substitutions to be made to transform a recipe ω into ω'. This operation is symmetric and we have:

$$d(\omega, \omega') = d(\omega', \omega) \quad (11)$$

In the same manner we may define the set N_δ of neighbours of a recipe $\omega \in \Omega$ defined as the number of configurations ω'. existing at distance δ from ω follows:

$$N_\delta(\omega) = \{\omega' \in \Omega \,|d(\omega, \omega') = \delta\} \quad (12)$$

The space in which it is possible to represent all the technological recipes through the reciprocal Hamming distance can be called *technological space.* The dimensionality of this space is given by number of neighbours $|N_\delta|$ for distance $\delta = 1$. Considering that each of the P instructions is characterized by S_{ij} values or choices, the dimensionality of the technological space would be:

$$|N_{\delta=1}| = \sum_{i=1}^{N} \sum_{j=1}^{M_i} (S_{ij} - 1) \quad (13)$$

Should the instructions have all the same number S of values or choices the, dimensionality of the technological space would become:

$$|N_{\delta=1}| = (S - 1)P \quad (14)$$

In this, case the geometrical representation of the technological space becomes a hypercube of dimension $|N_{\delta=1}|$.

A1.3 SPACE OF TECHNOLOGIES

Technological space is useful to describe a single technology with a defined operations structure representing all the configurations or recipes that this technology can assume following its model. When discussing various technologies, for example, studying technological competition and evolution, it may be useful to have a representation space for all technologies. This representation can be obtained by considering a family of technologies defined as able to fulfil the same specific human purpose. In order to describe the space of a family of technologies it is necessary to define a distance among the various technologies taken into consideration. Technologies cannot be described by a simple combination of operations because they also have a time-oriented structure that can be represented by a graph, and a graph can be mathematically represented in the form of a matrix. Distances among technologies can be then defined in terms of Hamming distances among matrices. Let us consider a set (family) of technologies T involved for the same human purpose, for example writing and transportation. Each technology belonging to T is characterized by M operations chosen from a set O of N different operations. This means that the same operations may be, in certain cases, repeated in the graph structure of a technology. Furthermore, some of the N operations can be also performed *in parallel*, i.e., at the same time. Hence, every technology $\tau \in T$ can be, associated with a M × N matrix T the elements, of which T_{ij}, can assume either value 1 or 0. More precisely, $T_{ij} = 1$ if the jth operations is present in the M position on the graph g related to τ, otherwise $T_{ij} = 0$. At this point, it is possible to establish a Hamming distance between any pair of technologies τ and τ' in T as the "difference" between their matrices T and T':

$$d(\tau, \tau') = \sum_{i=1}^{M} \sum_{j=1}^{N} |T_{ij} - T'_{ij}| \tag{15}$$

By knowing all distances among the technologies of the family T we may build up, as in the case of technological recipes, a space that we may name *space of technologies.* Furthermore, it is possible to define a set N_δ of the neighbouring technologies of the set T that are between the distance δ as:

$$N_\delta(\tau) = \{\tau' \in T | d(\tau, \tau') = \delta\} \tag{16}$$

The number of all the technologies τ present in a given family T is not univocally determined because it depends both on the type and on the "parallel" compatibility of the N operations. If, for instance, none of the N operations could be performed at the same time as another one in O, the cardinality of T would be simply given by N^M.

In the space of technologies, the Hamming distance between two technologies may be used as definition of the *radical degree* of a new technology as a measure of the difference between a new technology and a pre-existing technology in competition. In other words, new technologies that are at a short Hamming distance may be considered as the result of evolutive or incremental innovations, while new technologies that are at a long distance in this space may be considered as drastic or radical innovations in the frame of a technological paradigm. Such trajectory, in the technology space defined by our model, may be seen as a path at short Hamming distances in periods of incremental innovations and transitions at high Hamming distances in the presence of a radical innovation of a technology. In our model, technological space and space of technologies represent the exploration spaces for the development of a technology innovation.

A1.4 EFFICIENCY OF TECHNOLOGIES AND TECHNOLOGICAL LANDSCAPE

Technology efficiency is a complex concept that is difficult to define quantitatively in univocal terms. Technology efficiency in terms of, for example, energy or abated pollutants. can be measured quantitatively only by defining its specific aspects. An important type of technology efficiency is the economic efficiency that can be measured, for example, as the inverse of unitary cost of production. Relations between two types of efficiency may be established; relations between the various types of efficiency with economic efficiency are particularly important. The efficiency of a technology is strictly dependent on the particular used recipe. Certain recipes may have practically zero or negative efficiency, but other recipes may have high efficiency and constitute an optimum. As previously reported, associating all recipes of the technological space with the corresponding value of efficiency, we obtain the mapping of this space. Indicating with Θ the corresponding value of efficiency to a specific recipe ω of set Ω:

$$\Theta : \omega \in \Omega -> R^{+} \tag{17}$$

This mapped space is called *technological landscape* and it is characteristic of the specific structure of operations and instructions constituting a technology and depending, of course, on the used definition of efficiency. Exploring a technological landscape, we find regions with recipes with nearly zero efficiency and other regions with recipes with high values up to optimum values of efficiency. A representation of a simple technological landscape was previously presented in Fig. 2.5.

The efficiency of a specific recipe is generally a function of the efficiency of the various operations constituting the technology. In our model, we consider it convenient to define operation efficiency or inefficiency in such a manner that the sum of single operation efficiency or inefficiency constitutes the global efficiency or inefficiency of the recipe, respectively. For example, the efficiency θ_i of operation o_i, depends on values or choices s_{ijk} of its instructions p_{ij} but also possibly on values or choices of instructions of other operations o_l, $l \neq i$. The total efficiency $\Theta(\omega)$ of the technology with configuration ω composed by N operations is given by:

$$\Theta(\omega) = \sum_{i=1}^{N} \theta_i(o_i, o_1) \tag{18}$$

This way of calculating the total efficiency of a recipe as the sum of efficiency values of single operations is easily made in cases of technical efficiency such as energy, purity, and pollution abatement. In the case of economic efficiency, if we define it as the inverse of cost of each operation, the Equation (15) is not valid, as the sum of the inverse of operational costs does not give the total economic efficiency that is seen as the inverse of the sum of all the costs of operations. In such a case, it may be preferable to directly use the cost of operations, the sum of which constitutes the total cost of a recipe and optimal conditions in the technological landscape constituted by the minimum cost. In such a case, the total economic efficiency $\Theta(\omega)$ of the technology with configuration ω composed by N operations would be given by:

$$\Theta(\omega) = 1 \Big/ \sum_{i=1}^{N} c_i(o_i, o_l) \tag{19}$$

The total cost C of the recipe would be given by:

$$C(\omega) = \sum_{i=1}^{N} c_i(o_i, o_l) \tag{20}$$

It should be noted that a different definition of efficiency of a recipe is also possible: average of the sum of efficiency of the single operations, i.e., considering the inverse of cost of each operation.

A1.5 INTRANALITY AND EXTERNALITY OF A TECHNOLOGY

We have previously seen that the efficiency of an operation may be a function of the values or choices made for the instructions of the operation, but possibly also by instructions of other operations existing in the recipe. This means if we modify values of parameters of an operation oi, the efficiency θ_i of operation oi would depend on values or choices s_{ijk} of its instructions p_{ij}. but possibly also on values or choices of instructions of other operations o_l, $l \neq i$. This fact is defined as *intranality* of a technology. Operations efficiency as well as technology efficiency can be also influenced by external factors or variables that constitute in our model the *externality* of the technology and that should be taken into account in our model. External variables may be constituted, for example, by raw materials characteristics, differences in types or composition of used products, or various requirements in quality or types of certifications that production should satisfy. As has been previously done in the case of instructions values or choices, we may take in consideration various parameters for external variables forming specific external configurations in which the technology should operate. Considering the set V composed by B external variables v_i:

$$V = \{v_i, i = 1, \ldots, B\} \tag{21}$$

Each external variable v_i is characterized by a set R_i of R_i specific parameters:

$$R_i = \left\{q_{ij}, i = 1, \ldots, B; j = 1, \ldots, R_i\right\} \tag{22}$$

Where q_{ij} represents the jth parameter associated with the ith external variable v_i. The total number Q of parameters characterising an externality is given by:

$$Q = \sum_{i=1}^{B} R_i \tag{23}$$

The parameter q_{ij} may assume a set F_{ij} of values or choices:

$$F_{ij} = \{f_{jik}, i = 1, \ldots, B; j = 1, \ldots, R_i; k = 1, \ldots, F_{ij}\} \tag{24}$$

In which F_{ij} indicates the cardinality of the set F_{ij}.

Considering a specific externality with a set of B variables corresponding to a total of Q parameters, we can define as specific externality the specific configuration γ obtained by attributing a specific value or choice to each of the Q parameters. The set Γ of all the possible configurations of an externality are given by:

$$\Gamma = F_{11} \times F_{12} \times \ldots \times F_{1R1} \times \ldots \times F_{BRB} \tag{25}$$

In other terms we have:

$$\Gamma = \left\{ \gamma_1, 1 = 1, \ldots, \prod_{i=1}^{B} \prod_{j=1}^{R_i} F_{ij} \right\} \tag{26}$$

the number of configurations $|\Gamma|$ will be given by:

$$|\Gamma| = \prod_{i=1}^{B} \prod_{j=1}^{R_i} F_{ij} \tag{27}$$

Should be $F_{ij} = F, i = 1, \ldots, B\,et\,j = 1, \ldots, R_i$ we have:

$$|\Gamma| = F^R \tag{28}$$

We may note that the number of configurations of external variables also corresponds to the number of technology landscapes existing for the technology operating under the influence of a defined configuration of external variables. Finally, it is important to consider the value G resulting by:

$$G = |\Gamma| \times |\Omega| \tag{29}$$

$|\Omega|$ represents number of possible recipes existing in the technology landscape and $|\Gamma|$the number of externality configurations generated by external variables. Then G represents all the possible global configurations of a technology that takes into account both the number of possible recipes and the number of configurations of external variables that influence the efficiency of technology. We may easily represent the intranality and externality of a technology by building up a matrix constituted by columns representing all the operations o_j, I = 1 to N of a technology and rows representing all the instructions p_{ijk} i = 1, …, N and j = 1, …, M_i of the technology and all considered external parameters q_{ij}, i = 1, .., B and j = 1, …, R_i then assuming for each position a value of 1 whether influence of the specific instruction or external variable on the efficiency of the specific operation exists or 0 otherwise:

$$
\begin{array}{l}
o_1 o_2 \ldots\ldots o_N \\
p_{11} \ldots\ldots\ldots\ldots\ldots\ldots \\
p_{12} \ldots\ldots\ldots\ldots\ldots\ldots \\
\ldots\ldots\ldots\ldots\ldots\ldots \\
p_{NMN} \ldots\ldots\ldots\ldots\ldots\ldots \\
q_{11} \ldots\ldots\ldots\ldots\ldots\ldots \\
q_{12} \ldots\ldots\ldots\ldots\ldots\ldots \\
\ldots\ldots\ldots\ldots\ldots\ldots \\
q_{BRB} \ldots\ldots\ldots\ldots\ldots\ldots
\end{array}
$$

This matrix corresponds to a simplified adjacent matrix of a tri-parted graph constituted by the subset of instructions, the subset of external parameters, and the subset of operations with arcs that are oriented exclusively from instructions and external parameters nodes to operations nodes. This graph represents the global interactions that exist for a technology. Graphs may appear completely connected or in the form of clusters playing an important role in modeling a technology and designing exploration of correspondent technology landscapes. Such graphs may find application, for example, in experimental planning for reduction of the number of necessary experiments.

A1.6 APPLICATION OF THE MATHEMATICAL MODEL.

This model may be taken into consideration for various types of applications in the field of technology innovation and R&D management. However, using the mathematical model, even for a relatively simple technology, it would be necessary to treat a great number of recipes that increase exponentially with the number of operations, parameters, and values and that should be represented in a multidimensional space or technology landscape that is fully describable only by complex mathematical tools. We present here an example consisting of the design of experiments for quality assurance of a technology, adopting a certain number of simplifications, but in measure to give an idea of how the mathematical model of technology may be used for an application.

A1.7 DESIGN OF EXPERIMENTS FOR QUALITY ASSURANCE

The considered technology, called RUVECO®, was developed and patented in the late 1990s by Ruvaris, a company born as a joint venture of six Italian manufacturers of valves and faucets, with the aim of providing a method for the elimination of lead contamination from drinking water that originated from valves and faucets made with brass containing lead. This metal can be eliminated from the surface of mechanical parts by a selective dissolution and RUVECO® technology consists of a process of leaching lead by a suitable bath composition. It is important for this process to eliminate lead from the surface efficiently in order to reduce contamination of water under specific levels, in compliance with the various existing regulations for the products. Treated brass components may have various shapes and alloy composition, and optimal conditions of treatment should be found for every specific case. For the implementation of a quality assurance program, a set of suitable experiments have been considered to determine the optimal conditions of treatment as a function of: the externality of the technology essentially constituted by the alloy composition; type of fabrication used in the production of the part; and shape of the part under treatment, as well as limits of contamination to comply with certification standards. Cost of the treatment is a function of treatment time and consumption of bath that is essentially related to the concentration of the de-leading agent, as well as to consumption of degreasing and neutralizing agents that should be used in treatment. Optimum conditions are then defined as the minor cost of treatment necessary to reduce lead contamination to a level complying with certification standards. Considering the range of working parameters for the process, it is possible to build up a technological space of recipes and define two types of correlated technological landscapes using economic efficiency (cost) and de-leading efficiency, respectively. The technical efficiency is represented by the loss of lead on the treated samples, measured as increase of lead concentration in the bath. Furthermore, the treated samples should be tested to verify the respective level of contamination, as well as complying norms of certification, and determine a set of recipes whose samples comply with standards. Recipes complying with standards that have a minimum cost constitute the optimal recipes for the technology.

A1.8 MODELING THE RUVECO® TECHNOLOGY

RUVECO® technology consists of a simplified view of three main operations in sequence in three different treatment baths, indicated as follows:

Operation A: degreasing of parts by a suitable agent.
Operation B: selective de-leading of part surfaces by a suitable agent.
Operation C: neutralization by sweeping off residual de-leading bath from the parts.

In Table A1.1, we report the various instructions related to the three operations of the technology. In Table A1.2, we report the selected values for instructions implied in the operations, calculating s as the number of values or choices for each instruction.

The number of recipes of the technological space corresponding to the chosen range of instructions may be easily calculated using Equation (9), reported in the mathematical description of this appendix:

$$|\Omega| = 2 \times 2 \times 2 \times 2 \times 5 \times 2 \times 2 \times 2 \times 2 = 2^8 \times 5 = 1280 \qquad (30)$$

TABLE A1.1

Operations and instructions implied by RUVECO® technology

Operations	Instructions	Instruction Symbol
Degreasing	Temperature	A-1
	Time	A-2
	Degreasing agent concentration	A-3
Deleading	Temperature	B-4
	Time	B-5
	Deleading agent concentration	B-6
	Bath stirring	B-7
	Positioning of components	B-8
Neutralization	Time	C-9

TABLE A1.2

Number s of values or choices for instructions

Instructions	Values or Choices	s
A-1	2 temperatures (40° and 50°C)	2
A-2	2 times (5 and 10 minutes)	2
A-3	2 degreasing agent concentrations (high and low)	2
B-4	2 temperatures (40° and 50°C)	2
B-5	5 times (5, 10, 15, 20, 30 minutes)	5
B-6	2 de-leading agent concentrations (high and low)	2
B-7	2 levels of bath stirring (strong and medium)	2
B-8	2 possible positioning of components	2
C-9	2 duration of neutralization (long and short)	2

It is also interesting to consider the intranality of the technology that is represented in Table A1.3 in which the existing interactions between instructions and operations efficiency are indicated by a cross.

In addition to intranality, we should also consider externality of the technology that may be composed of four external variables, each characterized in our case by only one parameter, that may influence the process:

Variable V-1: Brass composition.
Variable V-2: Fabrication (wrought or cast component).
Variable V-3: Shape of the component.
Variable V-4: Certification (maximum admitted lead contamination).

TABLE A1.3

Intranality of RUVECO® technology

	Operations		
Instruction	**Degreasing (A)**	**Deleading (B)**	**Neutralisation (C)**
A-1	X	X	
A-2	X	X	X
A-3	X	X	X
B-1		X	X
B-2		X	X
B-3		X	X
B4		X	X
B-5		X	X
B-6		X	X
B-7		X	X
B-8		X	X
C-9			X

Choices made for external variables are reported in Table A1.4 indicating as f the number of values or choices.

The number of external configurations may be easily calculated by use of Equation (24), reported in in the previous part of this appendix and in the data of Table A1.4:

$$|\Gamma| = 3 \times 2 \times 2 \times 2 = 24 \tag{31}$$

There are 24 external configurations corresponding to 24 possible technological landscapes for each type of efficiency under consideration. The various external configurations also influence efficiency of operation and this externality is represented in Table A1.5 by indicating the existence of an interaction by a cross.

TABLE A1.4

Values or choices of external variables

External variable	Values or choices	f
V-1	3 alloy compositions	3
V-2	2 types of fabrications (wrought or cast component)	2
V-3	2 types of geometry (simple or complex)	2
V-4	2 types of certifications (easy or difficult)	2

TABLE A1.5
Externality of RUVECO® technology

	Operations		
External variable	**Degreasing (A)**	**Deleading (B)**	**Neutralization (C)**
V-1		X	
V-2	X	X	
V-3	X	X	
V-4		X	

Adopting such a technology model, it is possible to calculate the total number of positions that exist in the 24 possible technological landscapes by using Equation (26) and values of Equations (27) and (28):

$$G = |\Omega| \times |\Gamma| = 1280 \times 24 = 30720 \tag{32}$$

giving a total of 30,720 measurements of efficiency to completely describe the 24 landscapes.

A1.9 MAPPING OF THE TECHNOLOGICAL LANDSCAPE

We have previously seen that complete characterization of the 12 technological landscapes needs a very high number of efficiency measurements. This number can be reduced by introducing some simplifications in the model that are induced by technical and scientific knowledge on the process. These simplifications should take account of parameters and interactions that might have a limited or negligible influence on the efficiency of the technology from the scientific or technical point of view. In this way, we make a sort of mapping of the landscape that isolates a limited region that could probably contain the optimal working conditions and be characterized by a much lower number of positions. In the case of RUVECO® technology, we may consider that efficiency of degreasing and neutralisation operations are essentially dependent only on time, using standard temperature, and concentrations of the agents. On the other hand, the efficiency of the de-leading operations may be essentially dependent on temperature, time, and de-leading agent concentrations that neglect bath stirring and system of positioning in the bath. With respect to instructions in Table A1.3 and simplifying them as cited previously the number of recipes becomes:

$$|\Omega| = 2 \times 2 \times 2 \times 2 \times 5 = 2^4 \times 5 = 80 \tag{33}$$

External variables may be also reduced by not taking into consideration geometry of the part and by testing only under conditions of the most difficult certification for determining the set of recipes complying with the standard.

TABLE A1.6

Intranality and externality of RUVECO® technology in the simplified model

	Operations		
Instruction/Variable	**Degreasing (A)**	**De-leading (B)**	**Neutralization (C)**
A-2	X	X	
B-4		X	X
B-5		X	X
B-6		X	X
C-9		X	X
V-1		X	
V-2		X	

Adopting these simplifications in the variables in Table A1.5, the number of external configurations becomes:

$$|\Gamma| = 3 \times 2 = 6 \tag{34}$$

This means that the total number of measurements to characterize the six technological landscapes is:

$$G = |\Omega| \times |\Gamma| = 80 \times 6 = 480 \tag{35}$$

Finally, the intranality and externality of the technology may be described by integrating the data of Tables A1.4 and A1.5 and adopting the cited simplifications. The interactions obtained are reported by a cross in Table A1.6.

In conclusion, following the simplified model, the design of experiments considers the measurement of de-leading efficiency of 80 recipes in 6 different external configurations for a total of 480 recipes and calculation of economic efficiency (cost) of the 80 recipes. The obtained de-leaded samples would be submitted to verification of their acceptability following the selected certification determining the set of recipes complying with this standard. Comparing the cost of treatments of the set of complying recipes for each configuration makes it possible to choose the more efficient recipe for each external configuration (technological landscape) that will correspond to the recipe with the lowest cost. The knowledge of optimal treatment recipes, as a function of the various characteristics of the part that should be de-leaded, determines reliable conditions for establishing a quality assurance program in the use of the technology.

Appendix 2
Mathematical Simulation of the R&D Process

The model of the R&D process is based on organization of knowledge and capital looping fluxes, represented in Figure 3.1, as well as a dynamic based on projects. This model may be roughly simulated mathematically. Although adopting many simplifications, the mathematical model may be useful in studying territorial development and relations between R&D investments and economic growth. *The mathematical model simulates R&D investments in terms of number of R&D projects, carried out and economic growth in terms of successful technologies intended as technologies having important returns of investment.* Knowledge is seen in the model as composed of a certain number of packages of information generated by either terminated or abandoned projects that constitute the amount of available RDK. Such knowledge is increased by a certain percentage representing information packages coming from external scientific, technical, or other types of information finally constituting the total available knowledge for the generation of proposals for R&D projects. An innovative idea is considered to result from a combination of a certain number of information packages. In this way, it is possible to make a combinatorial calculation of the total number of potential innovative ideas based on the total number of available information packages. Of course, many such pure combinatory ideas would be absurd, but the number of innovative ideas that become R&D project proposals out of the total potential ideas, may be defined as the *innovative system efficiency* (ISE) of the territory that operates by a more or less advanced regime of open innovation and technology innovation carried out following a more or less advanced distributed innovation system. This means that the actors carrying out R&D projects are not limited to industrial R&D laboratories, but also to, for example, contract research laboratories, private or public research laboratories, start-ups, exchanging in a certain measure the generated RDK in the territory. Although normally a new technology is generated by a sequence of R&D projects, for simplification, in our mathematical simulation we consider that a new technology may be generated by a single project. Actually, generation of R&D project proposals, starting new R&D projects and formation of new technologies are continuous processes. However, in our mathematical simulation, we consider the R&D activity in term of cycles for simplification of computation; each cycle is fed by a certain number of R&D projects, generating or not generating new technologies. This means that duration of generation of R&D project proposals followed by carrying out of R&D projects is the same for all projects and equal to the cycle time. As RDK may be partly lost with time by a fading effect, we assume that a percentage of loss of knowledge occurs at each cycle and concerns all remaining knowledge of previous cycles. Calculations in the model are made considering a certain number of selection rates concerning: the rate of accepted R&D project proposals for financing, the rate of generation of new technologies from R&D projects, and the rate of new technologies that become successful technologies. Adopted rates are indicative and are based on R&D experience. Model calculation considers a certain number of initial financed R&D projects, and then computes the number of generated initial new technologies and the number of generated information packages.

This number of generated packages is reduced by fading effects for calculations in cycles successive to the first cycle. The total number of available information packages is obtained by adding a certain amount of external knowledge packages. Successful technologies are calculated from the number of generated new technologies by use of a suitable rate. With the total number of available information packages, it is possible, through a combinatory calculation, to obtain the number of generated R&D projects proposals by adopting various chosen values of ISE. Selection of proposals gives the number of financed R&D projects starting the activity of the successive cycle. For the definition of parameters of the model, we may separate those concerning the flux of capital from those concerning the flux of knowledge. In the case of flux of capital instead of return of investments, we have simply considered the generated number of new and possibly successful technologies using two rates, in terms of percentage of R&D projects that generates new technologies and percentage of new technologies that are successful. The mathematical structure of the model is the following:

Considering N the number of R&D projects carried out in a cycle, the number T of new technologies entering in use is determined by a selection rate v following the formula:

$$\mathrm{T} = v\mathrm{N} \tag{1}$$

Considering now the successful number of new technologies S, they are the result of a selection rate r on the number T of new technologies entering in use following the formula:

$$\mathrm{S} = r\mathrm{T} \tag{2}$$

Concerning the flux of knowledge, we first define a measure of knowledge generated by R&D projects in terms of number of information packages. For this purpose, we consider that each R&D project generates an average number p of information packages and that total available information packages result from the sum of packages generated by the cycle plus the information packages of previous cycles reduced by the fading effect f. Such total numbers of packages should be increased by a contribution taking account of information packages coming from scientific, technical, and other information composing an external available knowledge. The total number of information packages $\mathrm{I_T}$

available for generation of innovative ideas and then R&D project proposals may be calculated mathematically by the formula:

$$I_T = \left(N_L p + \sum_{i=1}^{n} I_i (1 - f) \right) (1 + E) \tag{3}$$

in which we have:

I_T: total number of information packages available for new innovative ideas after the last cycle

N_L: number of R&D projects in the last considered cycle

p: average number of information packages of each R&D project

n: number of past cycles

I_i: number of remaining information packages of past cycles from i = 1 to i = n

f: rate of fading effect (*)

E: fraction of added information packages by external knowledge

(*) It shall be noted that for remaining information packages of past cycles, we intend that the initial information packages of a cycle to be reduced by fading effect f at each successive cycle before the last one. With $f = 0$ the fading effect is not present, and with $f = 1$ there is a complete loss of past information packages.

The generation of potential innovative ideas is obtained by a combinatory calculation considering an average number of available information packages and number of combining information packages necessary to have an innovative idea and expressed by the following formula:

$$G = I_T (I_T - 1)/m \tag{4}$$

in which we have:

G: number of potential ideas for innovations

I_T: total number of information packages available for potential innovative ideas after the last cycle

m: combinatory number of information packages necessary to generate a potential innovative idea

In fact, such number G of potential ideas are a simple combinatory result, not considering any validity about specific combinations and necessarily contains, in fact, a large number of invalid or even absurd combinations. It is the task of the territorial innovative system to make a selection of valid innovative ideas. The number P of effective new ideas becoming R&D research proposals may be obtained by considering a rate factor s applied to the number G of potential new innovative ideas, such rate represents a measure of the *innovative system efficiency* (ISE) of a territory already cited, obtaining the relation:

$$P = sG \tag{5}$$

A last selection occurs in comparing R&D project proposals budgets with available R&D investments and we may define a rate t determining the number N of R&D proposals that can effectively become R&D projects following the relation:

$$N = tP \tag{6}$$

In conclusion, we may express the total number N of R&D projects carried out in a cycle as a function of generated packages of information I_T by the previous cycle combining Equations (4), (5) and (6):

$$N = tsI_T(I_T - 1)/m \tag{7}$$

However, in simplifying our application, we consider subsequently in running the model that all generated R&D proposals are valid and there is always enough R&D investment for the corresponding R&D projects. This means that we always considers $t = 1$ and consequently the number N of R&D projects will be equal to the number P of generated R&D proposals.

Considering the adopted functioning of the model with its simplifications, we may expect the formation of three scenarios. The first one corresponds to the case of introduction of a limited number of initial R&D projects and a low ISE and consequently, the number of R&D projects would be statistically insufficient for generating a new technology entering into use. On the other hand, a low efficiency in exploiting RDK may lead to a number of new proposals that are inferior to the initial number of R&D projects, decreasing the number of projects with the number of cycles; if that is not compensated sufficiently by available past knowledge, it would lead to a situation of decline and abandonment of the R&D activity from lack of generated proposals. In the second case, there is a sufficient number of initial R&D projects and acceptable ISE with an increase of the available RDK; however, the number of generated new technologies, after a reasonable number of cycles, may be statistically insufficient for at least one successful technology to implement the socio-economic growth of the territory. In this case, the technology evolves without assuring a real development, eventually entering into a stagnation phase, typically the Red Queen regime. In a third case, there is a sufficient number of initial R&D projects and a good value of ISE originating a reasonable high number of new technologies, and possibly successful technologies achieving the socio-economic growth of the territory. In this last case, RDK increases rapidly with the sequence of cycles and then the number of financed R&D projects becomes potentially enormous, although in reality, it will be limited by actual availability of R&D investments or by available human resources and structures for the R&D activity in the territory.

A2.1 PARAMETERS OF THE MODEL

In order to perform calculations, it is necessary to establish the value of the various parameters usable for the model calculations. As noted previously, most of the data necessary for parameters are practically not statistically available and we might consider only reasonable indicative estimations made possible by experience in R&D activity. The only parameter that might have a relation with statistical studies concerns the rate *r* determining the success of a technology following Equation (2). In fact, we have a study about value of patents carried out by Scherer and Haroff [Ref. 16 Chapter 4] showing a skew distribution of patents with a high values and, on the basis of results of such studies, we have assumed that only about 20% of new technologies (patents) result in substantial outcomes, indicating a value of 0.2 for the parameter *r* selecting the number of successful new technologies. Much more difficult is the estimation of parameter *v* determining the number of new technologies entering in use with respect to the number N of R&D projects carried out in a cycle following Equation (1). The rate of success of R&D projects becoming a new technology is quite variable depending, besides socio-economic factors, on the radical degree of the new technology while rate of abandonment of R&D projects differs, depending on the reached phase of the innovation process. On the other hand, an R&D project concerning an innovation with a limited novelty or low radical degree may have a higher probability of becoming a used technology. Taking account of previous considerations and experience in R&D, we might indicate a number of 40 R&D projects necessary to obtain a new technology entering into use and then an indicative value for parameter *v* of 0.025, which, however, should be considered the most uncertain parameter value for the model. The assigned parameters of *r* and *v* are set as constant in the studied application of the model, which means that on average, a total number of 200 R&D projects are necessary to generate five new technologies entering into use and only one of these five would be successful triggering a sensible positive socio-economic impact on the territory. Regarding parameters concerning the flux of knowledge of the R&D model, we have already defined the measure of knowledge in terms of number of information packages circulating in the flux. Quantitative data on generation of information packages by R&D projects and number of packages necessary for the combinatory calculation of innovative ideas are not available, but experience in R&D indicates that they cannot be a very high number for a single project. For this reason, we have considered, from a conservative view, a number of three information packages for parameter *p* associated to each project, and for parameter *m* an average number of two for information packages necessary for the generation of innovative ideas. Another parameter necessary for calculation concerns the fading effect on information packages generated in past cycles and including past external information. It has been considered that about 50% of past information is lost at each cycle. This means the total number of past information packages is halved at each cycle. The fading effect is then established to a value of 0.5 for the parameter *f*. The external contribution of information packages coming from scientific, technical, or other types of information to the total information packages available for generation of R&D projects cannot be, by experience, very high with respect to RDK. For this reason, we suggest for the external contribution an indicative added value of 10% of the total information packages generated by RDK, establishing a value of 0.1 for parameter *E*. Finally, there are two variables that are used for the parametric study of the model application that concern the initial number N_o of R&D projects and the rate of selection of innovative ideas becoming R&D proposals and then R&D projects. Concerning N_o, we have considered an initial number in a range from 10 to 100 projects. Considering the rate of selection *s* of potential combinatory number of ideas G, we have established a range between 0.1% and 1% of the percentage of combinatory ideas that are valid for R&D proposals. This means that we have a range between 0.001 and 0.01 for parameter *s*. In Table A2.1 we have summarized the values and ranges of the various parameters used for running the model.

The variable parameters calculated in the simulation model are:

I_T: total number of information packages available for starting a cycle
G: total number of potential combinatory ideas
P: total number of R&D project proposals
N: total number of R&D projects carried out in a cycle
T: number of new technologies entering in use
S: number of new successful technologies

Finally, we have considered a maximum number of cycles characterizing the effects of introduction of N_o initial R&D projects. This number has been established as 5 cycles.

A2.2 RESULTS OF THE MODEL CALCULATION

Before presenting results of calculations with the model, we stress the fact that this model is not a real reproduction of the complex activity of R&D, but only a simulation in which quantitative results are only indicative, depending on choice of parameters values that, in fact, do not result from any real statistical data but only from reasonable values suggested by experience. However, although the adopted simplifications, the model may give an idea of generation or selection processes occurring in a real R&D activity according to experience.

TABLE A2.1

Parameters used for model calculations and their values and ranges

Parameters	Value/range
r rate of success of technologies	0.2
v rate of generation of new technologies	0.025
t rate or R&D proposals selected for R&D projects	1
p number of information packages per R&D project	3
m combinatory number of information packages	2
f rate of fading effect	0.5
E rate of external information contribute	0.1
N_o number of initial R&D projects (range)	10–100
s rate of selection of innovative combinatory ideas for proposals (ISE range)	0.001–0.01

Model calculations have been simply implemented using an EXCEL® sheet. The calculation involves introducing an initial number of R&D projects for the first cycle and calculating the number of information packages formed by R&D. This number, increased with external information, gives the total number of information packages available for generation of R&D project proposals with numbers that depend on the adopted ISE rate *s*. As previously cited, we consider in this application a full availability of R&D investments for proposals and a number of R&D projects equal to the number of proposals. Should the number of R&D projects be high enough, there would be the generation of new technologies following the respective adopted rate *v*, and possibly successful technologies following the respective adopted rate *r*. The number of R&D projects generates a further number of information packages, increased by past-generated packages, reduced by fading effect, and increased by external information to give a new number of total information packages that allow starting calculations for the second cycle. A total of five cycles have been used to evaluate the effect of the initial number of R&D projects and the adopted ISE rate *s*. An example of the model as appearing in an EXCEL® sheet is reported in Figure A2.1 and it reports calculations results at the fifth cycle, using an initial number of R&D projects N_0 equal to 24, and obtaining a single successful technology in an innovation system with an efficiency value s = 0.005 (0.5%).

In a first run of calculations, we have determined the minimum number of initial R&D projects that are able to generate a number of R&D projects equal to the initial one following the adopted ISE value. Obtained results are reported in Figure A2.2. We may observe that in the case of the lowest ISE value equal to 0.01%, the number of initial R&D projects is very high (almost 200 projects) to obtain the expected result. This number may be considered excessive and unrealistic for a weak innovative system. In fact, using a low initial number of projects would result, in this case, in a continuous decrease of generation of new R&D projects, eventually terminating the R&D activity. For this reason, we have considered only an ISE value of at least 0.25% to a maximum of 1% for further calculations. Looking at Figure A2.2, the curve in fact separates the diagram area in two zones. Below the curve, the points represent a situation of technology stagnation and possibly decline, while above the curve, the points represent a situation of technology development but not necessarily of socio-economic growth if none of the new technologies becomes successful technologies. It should be noted that, in certain conditions, although an initial lower number of generated R&D projects exists, this number may become higher, after a certain number of cycles, because of accumulated past information, although reduced by the fading effect. In a second run of calculations we have determined the minimum number of initial R&D projects necessary to obtain at least one successful technology within a maximum of 5 cycles as a function of ISE values. Results are reported in Figure A2.3, and in this case the curve also separates two zones, the area above the curve representing a space with points corresponding to economic growth, i.e. generation of successful new technologies, and the area below the curve represents points corresponding to economy decline or stagnation because of possible formation of new but unsuccessful technologies. Using data obtained in these runs,

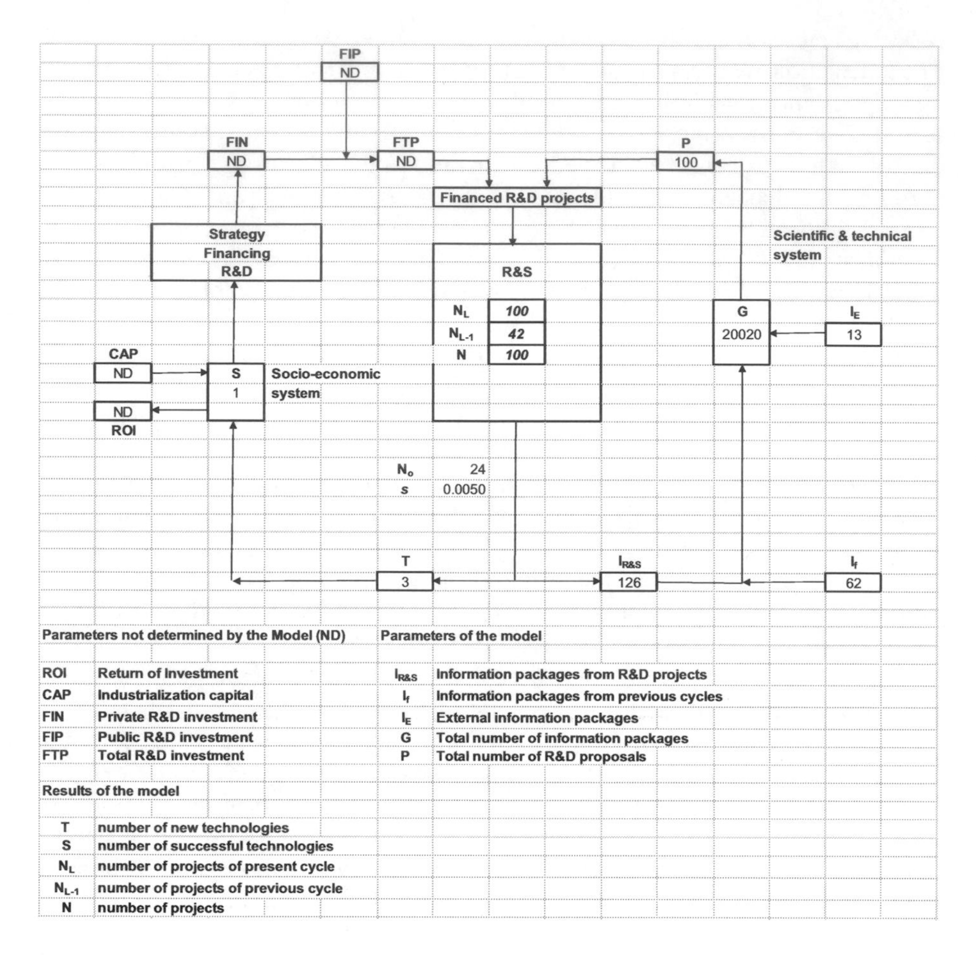

FIGURE A2.1 Example of Application of the R&D Model on an Excel® Sheet.

we may determine the minimum number of initial projects that gives the generation of at least one new technology and at least one successful technology, respectively, as a function of adopted ISE after a maximum of five cycles. The obtained two curves are reported in Figure A2.4. In this figure there are three areas: the first one, above the higher curve, represents a zone of points corresponding to generation of successful technologies and then of growth; the second one, below the lower curve represents a zone with points corresponding to absence of generation of new technologies and then of decline; the third one, between the two curves represents a zone with points corresponding to generation of new but unsuccessful technologies and consequently, without any real influence on growth. This last case corresponds to the presence of new technologies but not of competitive technological advantages characterizing successful technologies, which is typical of a Red Queen regime. Finally, in Figure A2.5 we report

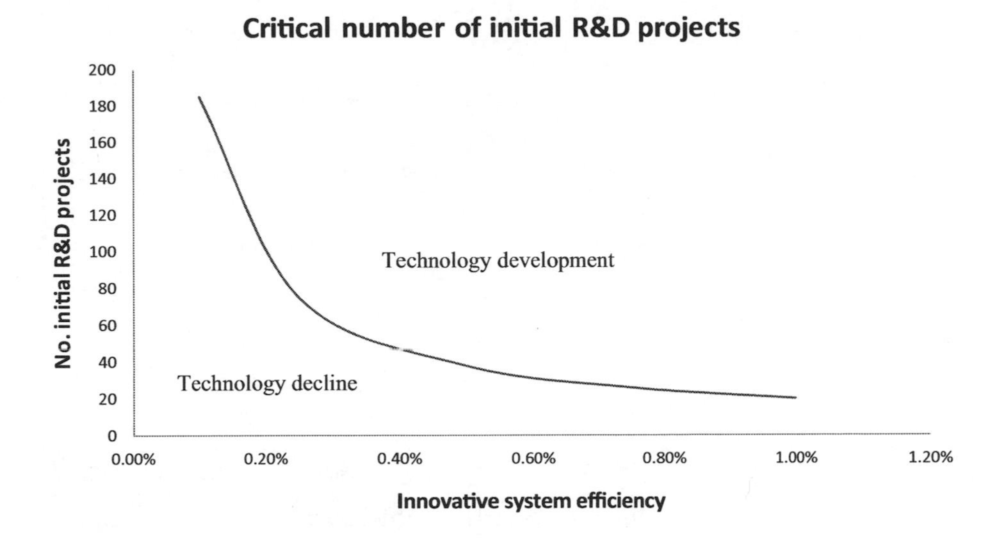

FIGURE A2.2 Minimum Number of Initial R&D Projects to Obtain the Same Initial Number.

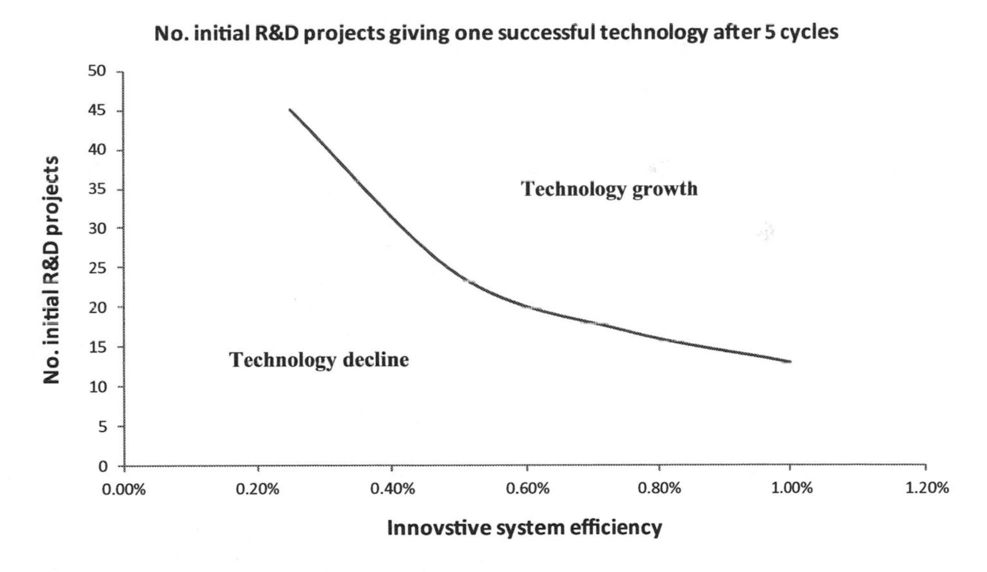

FIGURE A2.3 Number of Initial R&D Projects Necessary to Obtain at Least One Successful Technology.

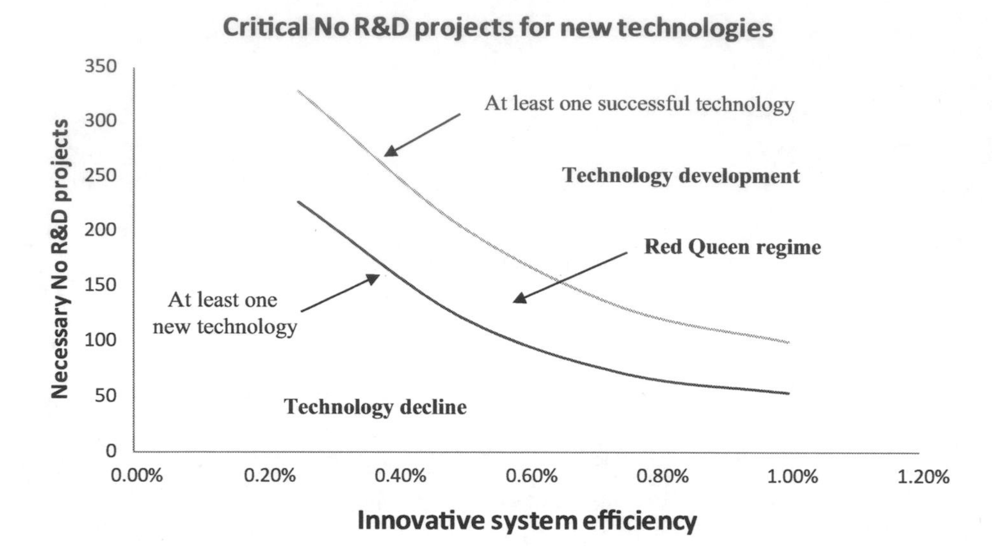

FIGURE A2.4 Various Areas Corresponding to Different Technological Regimes.

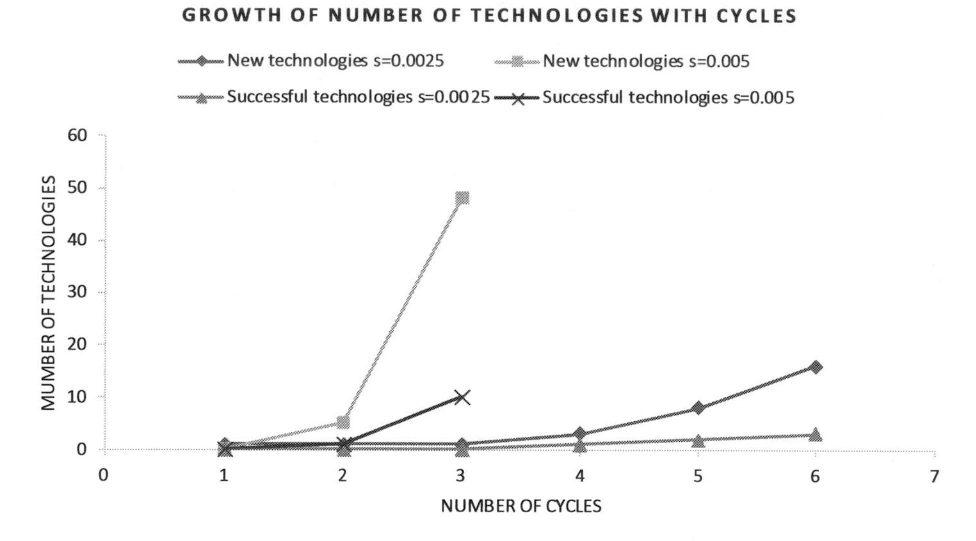

FIGURE A2.5 New and Successful Technologies as a Function of Number of Successive Cycles.

results of a third run about the evolution of the cumulated number of new and successful technologies as a function of the number of successive cycles. For this calculation, we have chosen an initial number of R&D projects equal to 50, and generation of new technologies and successful technologies using two intermediate values of ISE, s=0.005 and s=0.0025, respectively, that correspond to a high level and a low level of innovation efficiency, respectively, in the territory. Reported results show easily that for high efficiency (s=0.005), there is an exponential growth in number of new technologies, and then of successful technologies, according to what it is expected by the R&D model when financing of R&D projects is not limited. For the high value of ISE, at the fourth cycle the calculated number of new technologies is 2828 and that of successful technologies is 566, not represented as out of scale. In calculations made with a low ISE (s=0.0025), there is, considering six cycles, only a limited increase in new technologies and a small linear increase in successful technologies. Of course, the extremely high growth of technologies in a territory, observed for high ISE, is not necessarily realistic, and limitations to the number of financed R&D projects could appear in the face of such high number of potential R&D projects; another limitation may be simply constituted by lack of structures or human resources to carry out such huge numbers of R&D projects.

A2.3 INFLUENCE OF PARAMETERS ON MODEL RESULTS

The influence of parameters on model results depends on how the model operates and, although it is only a rough simulation of R&D activity, the model calculations might suggest interesting observations about the real functioning of the R&D process. In fact, the model operates through a sequence of generative and selective process steps. Generative steps concern the combinatory process among information packages, constituted by RDK and external information, while the number of available R&D project proposals is determined by the innovation system efficiency ISE of the territory. Selective parameters determine the number of financed R&D proposals, the number of R&D projects generating new technologies, and the number of successful technologies appearing among the generated new technologies. All selective parameters induce a linear dependence between number of projects proposals and the resulting number of technologies, while the generation process of innovative ideas induces an exponential dependence because of the combinatory process at the base of calculations. In fact, the presence of selective parameters results in a critical minimum number of projects in order to reach the conditions for generation of new technologies, and a higher critical number to obtain successful technologies. On the other hand, the combinatory process makes the generation of R&D projects an autocatalytic process that favors an increasing growth of new and possibly successful technologies. Experience in R&D activities demonstrates the validity of such results; in fact, it is well known that low investments in R&D produce poor values of growth, while effective territorial innovation systems, combined with full availability of R&D investments, are able to produce exponential growth of technologies and positive socio-economic impacts, as observed, for example, in the case of Silicon Valley.

Appendix 3
Mathematical Simulation of the SVC Cycle

The activity of the SVC system is of a cyclic nature that is presented in Figure 3.4. The cycle starts with a certain number of start-up projects that are submitted to VC for financing. Part of these proposals are accepted, the remaining rejected. Accepted start-ups are financed and, during the development part of start-ups, is abandoned; the remaining part reaches an exit that produces a return of investment (ROI) to VC. Part of this return remains as margin to the VC and the rest of the capital is reinvested for financing new start-ups. The equilibrium condition of this system is reached when reinvested capital in new start-ups is the same as the capital invested at the beginning of the cycle. This also means that the return of investment should be higher than this initial capital to compensate the return of the venture capital that is used to cover its operative cost and a possible reward. In the activity of the SVC cycle, we may distinguish two important strategies concerning the selection of financed start-ups as follows.

Strategy A. In this case, for selection, the potential return of investment of the start-up and the capacity and experience of the team are primarily considered. The feasibility of the start-up is taken into account as well, but is considered too uncertain to be the main criterium of choice; the financing of a high number of start-ups is preferred, in order to increase the probability of start-up exits that reach a great return of investment.

Strategy B. In this case, the feasibility of the start-up is of main importance in the choice, and suitable methods for its determination are taken into consideration in order to reduce losses due to abandoned start-ups. Experience of the team does not have the same importance as in Strategy A, and in certain cases, the failure of a team in a previous start-up is considered the result of bad management instead of available experience in start-up management.

These two strategies may be studied in the SVC system by developing a mathematical simulation based on the cycle reported in Figure 3.4. In the simulation it is considered the case of a territory in which are operating a certain number of VC companies financing start-ups and in which, at the beginning of a cycle, are presented a certain number of start-up project proposals. These proposals may be selected following Strategy A or B. The number of chosen start-ups for financing would be higher in Strategy A than in Strategy B. On the other hand, the number of abandoned start-ups would be lower in Strategy B than in Strategy A as it may be deduced by the differences in the types of selection in the two strategies. As previously cited, both strategies have a point of equilibrium in which the ROI, after deduction of margin retained by the VC, reaches the same amount of capital invested in the start-ups at the beginning of the cycle. From the description given previously, Strategy A, with a higher number of abandoned start-ups, should have a greater average ROI for start-ups than Strategy B, in order to reach equilibrium. Such situations may be determined by the ratio between the amount of obtained ROI and the total invested capital that satisfy the condition of equilibrium, i.e. the availability of capital, after deduction of margins retained by VC, equal to the initial invested capital. Starting from this condition of equilibrium, it also possible to carry out parametric studies calculating the effects of changes in the rate of accepted start-ups for financing, the rate of abandonment of start-ups, and the effects of increasing the average ROI on the availability of capitals for financing new start-ups.

A3.1 CALCULATION AND RESULTS OF THE MATHEMATICAL SIMULATION MODEL

Calculations have been made using an Excel® sheet and an example is reported in Figure A3.1. Considering a first cycle with an initial number P of start-up projects the number of accepted proposals S is be given by:

$$S = sP \tag{1}$$

in which s represents the rate of accepted projects. In the simulation in Strategy A, it is assumed that there is always full financing of all proposed projects and

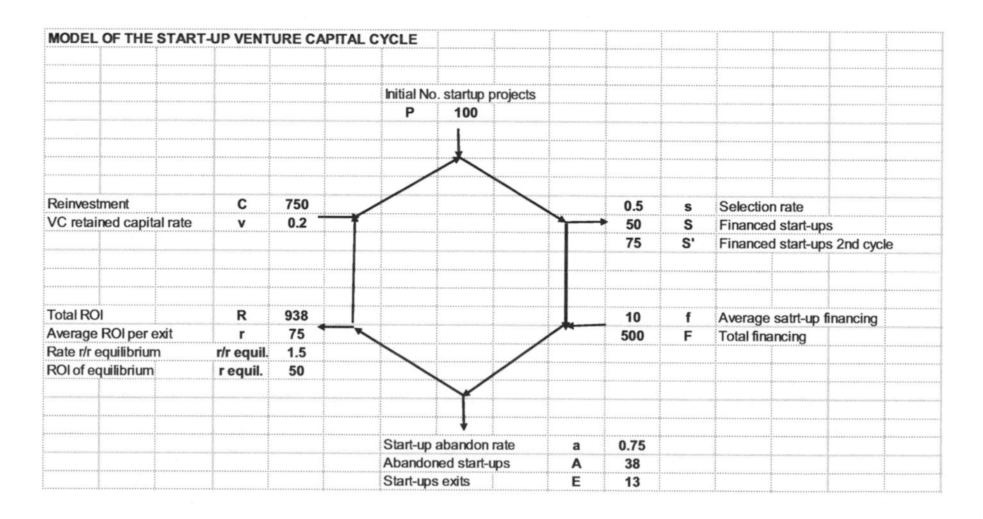

FIGURE A3.1 Example of Application of the SVC Cycle in an Excel® Sheet.

that s would be equal to 1. In, Strategy B the number of accepted proposals would be lower and s may assume various values that, in our calculation, correspond to various percentages of acceptance, for example, the range between 25% and 75% corresponds to s varying between 0.25 to 0.75 (0.25, 0.35, 0.45, 0.55, 0.65, 0.75). The total investment made by VC is composed of the sum of investments made for either abandoned or successful start-ups, reasonably presenting in the two cases a difference in the average capital financing. Investments may be expected to be lower for abandoned start-ups than for start-ups that reach a positive exit and achieve full development. In our simulation, we have in fact, made a simplification considering an average f of capital spending for both successful and abandoned start-ups, assuming that the differences might not influence in to a great extent the results of the simulation. Assuming that the average financing of each start-up is f, independently of whether it is abandoned or not, the total financing of the venture capital F would be:

$$\mathrm{F} = f\mathrm{S} \tag{2}$$

Considering now the rate of abandonment of the S financed start-ups, the situation is different based on the two strategies. It is reasonably expected that the abandoning rate a in the case of Strategy A would be much higher than in the case of Strategy B. In fact, in territories where Strategy A is prevalent, as in the Silicon Valley, the rate of abandonment is currently estimated around 90% i.e. a is equal to 0.9. In other territories, where the Strategy B is prevalent, the circulating value of rate of abandonment is between 70% and 80%, and in our calculation, a has been established with a value of 0.75. The number A of abandoned start-ups would be:

$$\mathrm{A} = a\mathrm{S} \tag{3}$$

and the number E of start-ups having a positive exit would be:

$$\mathrm{E} = (1 - u)\mathrm{S} \tag{4}$$

Considering now an average value r of the return of investment of start-ups having a positive exit, the total return of investment R is expressed as:

$$\mathrm{R} = r\mathrm{E} \tag{5}$$

Consequently, in our calculation, we assume that F represents the total capital expenditure in financing start-ups that should be covered at equilibrium by the return of investment R. In fact, we have to consider that a part v of return of investment R would remain to the venture capital to cover its operational cost and possible margins. Such value is estimated as 20% or v equal to 0.2 and, in this case, the amount of capital C that would be available for reinvestment in start-ups would be expressed as:

$$C = R(1 - v) \quad (6)$$

In this way, the number of start-ups S that may be financed in a new second cycle would be expressed as:

$$S = C/f \quad (7)$$

Assuming the same average value f for financing of the previous cycle, the equilibrium point of the cycle may be also considered as reached when the value of C makes possible the same number of financed start-ups in the new cycle as in the previous one, supposing the same financing average. The point of equilibrium may be obtained by iterative calculations based on varying the value of the average return of investment r of successful start-ups. The r value of equilibrium would, of course, be different in the case of Strategy A or Strategy B. Calculations have been made that consider the following starting parameters for the two strategies:

Strategy A
Rate of financing $s = 1$
Rate of abandoning $a = 0.9$
Strategy B
Rate of financing s variable from 0.25 to 0.75
Rate of abandoning $a = 0.75$

For the calculation using Excel® we have considered an initial number of 100 start-up projects and an average investment for each financed start-up equal to a relative reference value of 10. The results obtained for the equilibrium are:

Strategy A
Average ROI per exit at the equilibrium: 125
Ratio r/f at the equilibrium: 12.5
Strategy B
Average ROI per exit at the equilibrium: 50 (found independent of the value of a)
Ratio r/f at the equilibrium: 5

It should be noted that, as expected, the ROI necessary for equilibrium in Strategy A is much higher than in the case of Strategy B (ratio of 12.5:5). Calculations also show the independence of ROI per exit by the value of acceptance s, an expected consequence of the proportional decrease in need for financing capital with the number of financed start-ups in the established condition of equilibrium. In fact, in the parametric study, taking constant the selection rate (s equal to 0.5) but varying the abandoning rate a from values of 0.9 to 0.6, the average ROI per exit for equilibrium varies from 125 to 31.5, and consequently the ratio r/f from 12.5 to 3.15.

The model may also be used for a parametric study that considers the calculation of the number of financed start-ups that may be obtained following the two strategies. This calculation starts from the value of equilibrium of both strategies, and then considers a growing ROI expressed by the increase of the rate of ROI on the ROI of equilibrium from value of 1.5 to 4.0. The results are reported in Figure A3.2 and it is observed that the number of new

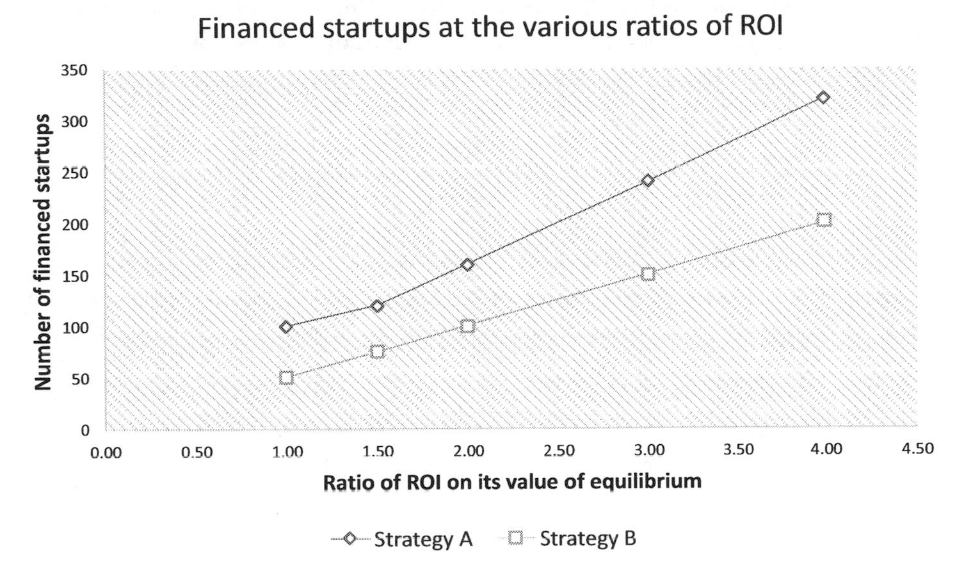

FIGURE A3.2 Number of Financed Start-ups Following the Ratio of ROI on ROI of Equilibrium.

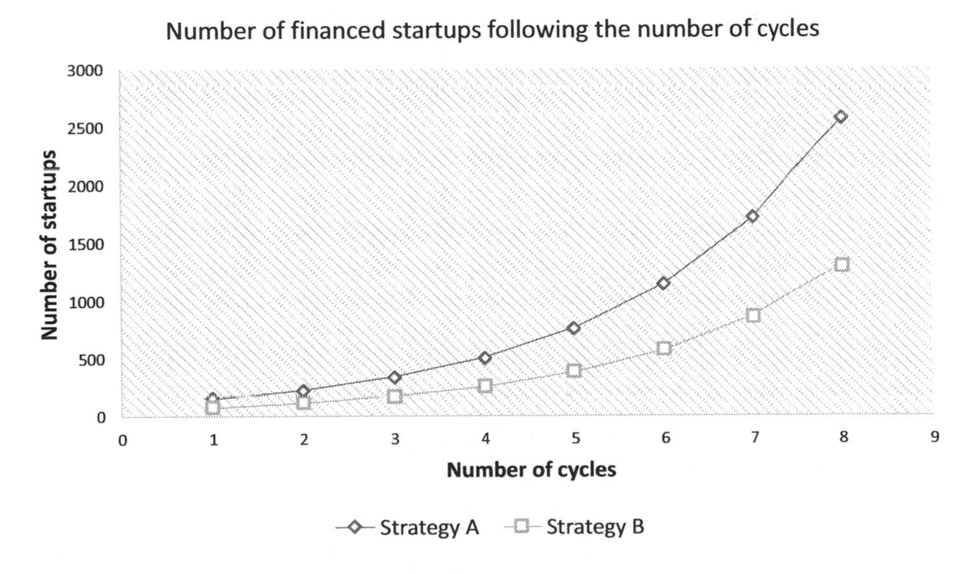

FIGURE A3.3 Number of Financed Start-ups as a Function of Number of Cycles.

financed start-ups increases proportionally with increase in the considered ratios, and that the increase is higher in the case of Strategy A than in Strategy B. A further parametric study calculates the number of new start-ups that may be financed by capital generated in a cycle as a function of the number of cycles following Strategies A and B. The results are reported in Figure A3.3 and show an increase in the number of financed start-ups in the case of Strategy A, confirming that this strategy has a higher capacity for financing than does Strategy B. Actually, the model calculations also show that the ratio of financed start-ups in Strategies A and B of each cycle remains the same as in the first cycle. Thus, the key factor that makes Strategy A more successful is not the consequence of the cycle, which presents, in fact, linear dependences following the adopted parameters rates, but by the fact that Strategy A plays with a higher number of financed start-ups with a high ROI potential that results in higher ROI values. Following the SVC model simulation, we reach the conclusion that the high rate of accepted start-up projects and the high availability of financing capitals, along with a good know-how in selection of start-ups with a high ROI potential for financing, are the basis of the success of Strategy A.

Index